SWIRLING FLOWS AND FLAMES

Edited by **Toufik Boushaki**

Swirling Flows and Flames
http://dx.doi.org/10.5772/intechopen.75295
Edited by Toufik Boushaki

Contributors

Sergey Alekseenko, Vladimir Dulin, Aleksei Lobasov, Dmitriy Markovich, Julio Marañón Di Leo, Juan Sebastián Delnero, Guillermo Capittini, Trevor Jones, Yuri Dubnishchev, Vitaly Arbuzov, Lukashov Vladimir, Konstanin Sharov, Vadim Lemanov, Shan Zhong, Yuanye Zhou, Toufik Boushaki

First published in London, United Kingdom, 2019 by IntechOpen
IntechOpen is the global imprint of INTECHOPEN LIMITED, registered in England and Wales, registration number: 11086078, The Shard, 25th floor, 32 London Bridge Street
London, SE19SG – United Kingdom
Printed in Croatia

British Library Cataloguing-in-Publication Data
A catalogue record for this book is available from the British Library

Additional hard and PDF copies can be obtained from orders@intechopen.com

Swirling Flows and Flames, Edited by Toufik Boushaki
p. cm.
Print ISBN 978-1-83880-743-6
Online ISBN 978-1-83880-744-3
eBook (PDF) ISBN 978-1-83880-745-0

For EU product safety concerns:
IN TECH d.o.o., Prolaz Marije Krucifikse Kozulić 3, 51000 Rijeka, Croatia,
info@intechopen.com or visit our website at intechopen.com.

Meet the editor

Dr. Toufik Boushaki is currently Associate Professor of Mechanical Engineering at the University of Orleans, France. He conducted research on combustion at the Institut de Combustion Aérothermique Réactivité Environnement (ICARE) laboratory of the National Center for Scientific Research (CNRS). He received his PhD in Energetics (2007) and DEA in Reacting Flows (2003) from the Institut National des Sciences Appliquées (INSA) of Rouen, France. Dr. Boushaki obtained an MSc in Mechanical Engineering from Blida University, Algeria (1999). His research interests focus on turbulent combustion, control of flames, and plasma and pollutant emissions. He is a member of the French section of the Combustion Institute.

Contents

Preface

A swirling flow is due to the application of a spiral movement. A tangential (or azimuthal) velocity component is attributed to the flow through blades (by swirlers) or the use of small tubes placed tangentially to the burner axis. Swirl flows are used in a wide range of industrial applications. In non-reacting cases, examples of applications include vortex amplifiers and reactors, heat exchangers, jet pumps, cyclone separators, and so on. In reacting cases, swirlers are widely used in combustion systems, such as gas turbines, industrial furnaces, boilers, gasoline and diesel engines, and many other practical heating devices. Effects of using swirl on flow and combustion are significant and varied, and concern, for example, aerodynamics, mixing, flame stability, intensity of combustion, and pollutant emissions.

This book collects recent research on swirling flows, in reacting and non-reacting cases, including in particular its presence in practical systems. The purpose of this book is to present recent research efforts to understand and characterize swirling flows of different types and in different applications. The book contains five chapters and a general introduction on the subject.

In the Introduction, the author firstly defines swirl flows, their advantages, and their different applications. Secondly, the author discusses swirl generation techniques and the swirl number that characterizes the intensity of rotating flows. Thirdly, the author reports the swirl effects on flames and pollutant emissions, including different studies from the literature. Chapter 2 presents results of coherent structures in the near field of swirling turbulent jets in reactive and non-reactive conditions. The authors are interested in analysis of these structures' impact on mixing of a passive tracer in the jets and on regular flame front deformations. They performed experiments with fuel-lean and fuel-rich methane/air mixtures in the case of different swirl numbers using particle image velocimetry/planar laser-induced fluorescence (PIV/PLIF) measurements and proper orthogonal decomposition (POD) of the velocity snapshots. Chapter 3 investigates the aerodynamic structures of flow in reactive and non-reactive cases. Two type of flow are studied: free jet and jet with a cross flow. Measurements techniques including visualization in the ultraviolet region, PIV, and hot-wire anemometry are used. Propane and hydrogen in a mixture with diluents (CO_2, He) are adopted as a fuel. In the reacting case a propane-butane mixture and Freon 22 is tested. The results concern flow visualizations, velocity fields, and turbulence intensities. Chapter 4 deals with the detection, generation, and characterization of vortexes. Two techniques to generate a vortex are used: a Vortex Generator (VG) with fixed surface, and with blowing air. The characterization of vortexes is based on some properties such as velocity field, vorticity field, and turbulence intensity. Some results are presented as velocity signals and distributions, CWT, and visualization of vortex of turbulent free flow. Chapter 5 investigates the flow of swirling liquid in a duct. The emphasis has been on Spanner-type profiled tubes, but this is by no means the only way to generate swirl. The efficacy of the first-order Solid Body Model is demonstrated by the simulation of flow through a 10.0 m cylindrical tube. The prediction that the downstream data taken after a distance of 3 tu would be representative of fully developed flow is amply demonstrated. Chapter 6 presents characteristics of particle adhesion patterns in a cyclone particle separator and their mechanisms based on experimental and simulation results. It also investigates the effect of particle inlet position on adhesion. In addition, the chapter studies 2D modelling of particle adhesion. The authors' key finding of the 2D modelling work is that electrostatic force plays an important role on the particle adhesion.

Dr. Toufik Boushaki
ICARE CNRS
University of Orléans, France

Chapter 1

Introductory Chapter: Swirling Flows and Flames

Toufik Boushaki

Additional information is available at the end of the chapter

http://dx.doi.org/10.5772/intechopen.86495

1. Introduction

Swirling flows are used in a very wide range of industrial applications. In non-reacting cases, examples of applications include vortex amplifiers and reactors, heat exchangers, jet pumps, cyclone separators, whirlpools, tornadoes, etc. In reacting cases, swirlers are widely used in combustion systems, such as gas turbines, industrial furnaces, boilers, gasoline and diesel engines and many other practical heating devices. Effects of using swirl on flow and combustion are significant and various and concern, for example, aerodynamics, mixing, flame stability, intensity of combustion and pollutant emissions. In this chapter, we are interested in the use of swirling flows in combustion systems.

In industrial plants, the geometric methods of flame stabilization are based on increasing the residence time of reactive gases either by the wake effect *bluff-body burner*, by the flow rotation of reactants *swirl burner* or by a combination of these two mechanisms.

Burners with helical flows generally referred to as the swirl burner have very wide applications in the industrial field. In the literature, many examples of uses of swirling jets are found as a means of controlling combustion [1–4]. Other studies focused on the swirl effect on the characteristics of lifted flames [5], stabilization and blow-out phenomenon [6–8] and pollutant emissions [9–13].

2. Swirl generation techniques

There are several ways to generate the rotation of a flow. They can be classified into three main categories:

- Use of fins or adjustable propellers tangentially deflecting the axial flow. Because of its simplicity, this device is generally used in industrial systems, in particular in gas turbines. However, this type of device introduces significant head losses, and the intensity of swirl is limited (design of fins) [14].
- Rotating mechanical devices which generate a rotational movement to the fluid passing between them [10].
- Tangential injection of part or all fluid quantity into a main duct. The intensity of the swirl is then determined by the ratio between the flow injected tangentially and that injected axially [15, 16].

3. Swirl number

The swirl number (S_n) is a dimensionless number that characterizes the rotating flows. It is defined as (Beér and Chigier [1])

$$S_n = \frac{G_\varphi}{RG_x^{'}} \tag{1}$$

where G_φ is the axial flux of the tangential momentum, $G_x^{'}$ is the axial flux of axial momentum and R is the exit radius of the burner nozzle. This number determines the intensity of swirl; the more this number is higher, the more the swirl effect is strong. The fluxes G_φ and G_x are given by the following expressions:

$$G_\varphi = \int_0^R (Wr)\ \rho U\ 2\ \pi\ r\ dr \tag{2}$$

$$G_x = \int_0^R U\rho\ U\ 2\pi\ r\ dr + \int_0^R p\ 2\pi\ r\ dr \tag{3}$$

U and W are the axial and tangential components of the velocity, respectively; p is the static pressure of the flow.

The introduction of this swirl number is based on G_φ and G_x measurements along the axis the flow. To experimentally evaluate S_n, it is necessary to have access to the velocity and pressure distributions on cross sections of the rotating flow. The swirl number can be calculated geometrically with the dimensions of the swirler and burner, as in [17, 18].

4. Swirl effects on flames

According to some studies as Beér and Chigier [1], for flows with a low swirl number ($S_n = 0.6$), there is no recirculation of the flow (**Figure 1**). The swirl induces an increase in the

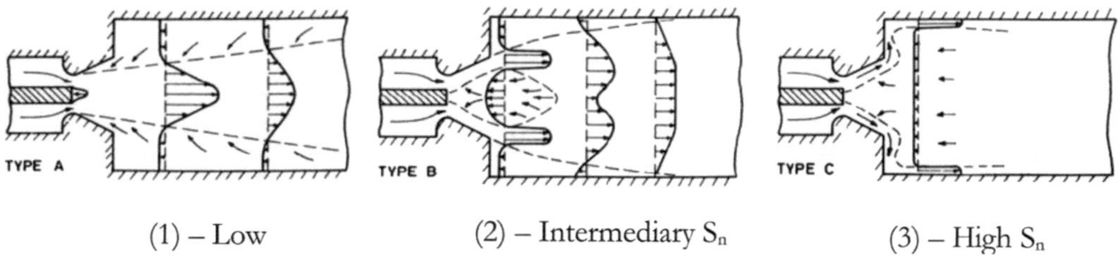

Figure 1. Swirl effect on flames [1]. (1) Low S_n, (2) intermediary S_n and (3) high S_n.

entrainment of the ambient fluid and a decrease in the axial velocity of the flow. Axial velocity radial profiles remain a Gaussian form as long as the swirl remains below 0.5. Velocity maximum deviate from the axis when approaching $S_n = 0.6$. Thus the degree of jet opening and the entrained mass flow then increase continuously with the swirl number. From a swirl intensity of about 0.6, a recirculation zone appeared in the main flow. The size and position of this recirculation zone varies with the intensity of swirl [1, 17]. This zone is an essential element for the stabilization of flames because it contains preheated fresh gases and allows a better combustion.

The swirl is commonly used in combustion because it has a stabilizing role for the flame [1, 8]. Indeed, if the swirl intensity is strong enough, a recirculation zone will appear. The latter is a region in which fresh gases and flue gases are well mixed. In addition, swirl improves the entrainment of gases and greatly increases flame-blow limits [3, 7]. Beer and Chigier [1] show the effect of the swirl intensity on the behaviour of the flame as shown in **Figure 1**.

- Case (1): The intensity of swirl is low; the flame behaviour is similar to that encountered without swirl. The flame is detached from the burner.
- Case (2): The intensity of swirl is intermediary; the flame is stabilized closer to the burner, in the recirculation zone rich in fresh gas. Turbulence levels are high and combustion is intense.
- Case (3): The intensity of swirl is high; the interaction of flame/walls is intense; this is in generally an undesirable case, except in certain industrial furnaces to obtain an intense and uniform radiation.

Many advantages of swirling reacting flows are now well-known and studied, as the flame stabilization improvement through the vortex breakdown phenomenon, due to the central recirculation zone (CRZ) occurrence [3, 7, 19, 20] and enhancement in turbulent mixing of reactants by means of the precessing vortex core (PVC). Authors [21–25] studied the flow characteristics and mixing properties of the swirling double concentric jets at low Reynolds numbers with control discs. Jourdaine et al. [26], investigated premixed swirling flames of CO_2–CH_4–O_2 and CH_4-air using different optical diagnostics. Benim et al. [27] studied numerically turbulent swirling flames in a model gas turbine combustor, using OpenFOAM code through the LES approach. Elbaz and Robert [16] investigated non-premixed methane swirling flames and particularly the quarl geometry effect on the flame structure. They were interested in the measurements of the turbulent flow field, concentrations of species and temperature distribution. Khalil Ahmed and Gupta [28] studied the colourless distributed combustion (CDC) combustor. CDC was examined with both swirling and non-swirling cases using different injection of velocity to improve distributed combustion. Orbay et al. [29] reported experiments and numerical simulations of swirling turbulent flows in non-reactive and reactive cases in an academic gas turbine combustor. They conducted PIV-LDV measurements and LES calculations focusing on the characterization of the flow structures, the effect of inflow swirl and the heat release in a premixed natural gas-air combustor. Candel et al. [30] performed a review of some recent progresses on swirling flame dynamics, particularly with the presence of acoustic perturbations. They found that the presence of acoustic perturbations in swirling flows induces a vorticity wave which is convected by the flow. They showed that the response of flame is characterized by a combination of heat release rate fluctuations generated by the incoming acoustic and convective perturbations. Von Lavante and Yao [31] studied numerically different types of axisymmetric internal configurations to examine the development of turbulent swirling and non-swirling flows along the flow. A numerical algorithm for solving the three-dimensional axisymmetric internal turbulent flows was developed. Good agreement of the calculation and theoretical results of the velocity profiles and pressure distributions was obtained. Iyogun et al. [14] presented an experimental investigation on the stability of swirling non-premixed CH_4 flames in a coaxial burner. Measurements were conducted for both reacting and non-reacting swirling flows. Cozzi and Coghe et al. [32] investigated by stereo-particle image velocimetry (S-PIV) technique the initial region of a swirling gas jet on a model burner in isothermal swirling air jets. Their study was interested to identify the presence of vortex breakdown, the central recirculation zone (CRZ) and the precessing vortex core (PVC). Boushaki et al. [13] investigated turbulent flames on a swirl burner with a radial injection of fuel (**Figure 2**). The burner consists of two concentric tubes with a swirler placed in an annular part supplying the air. Eight vanes are made to induce a swirl intensity variation. The central pipe delivers a fuel radially through eight holes symmetrically distributed on the periphery of the tube near the exit burner.

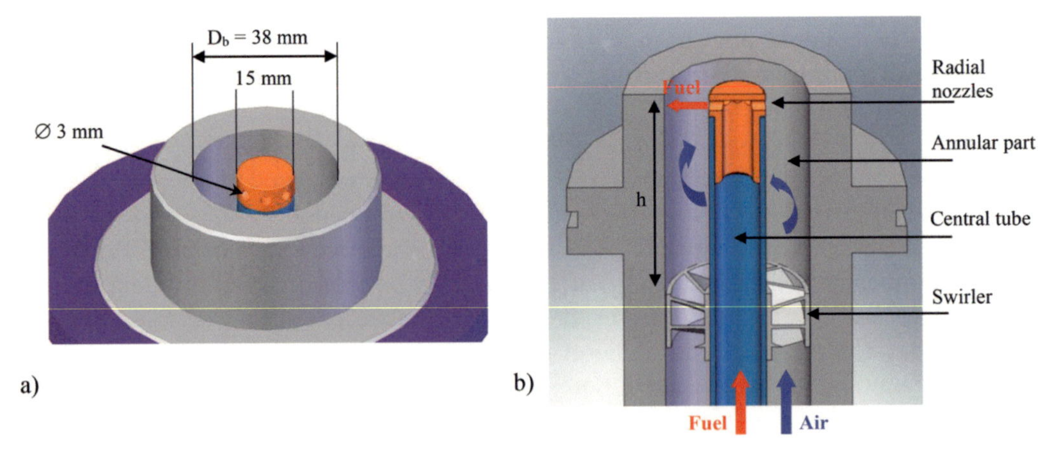

Figure 2. Diagram of the coaxial swirl burner, (a) 3D view of the top part and (b) vertical cross section.

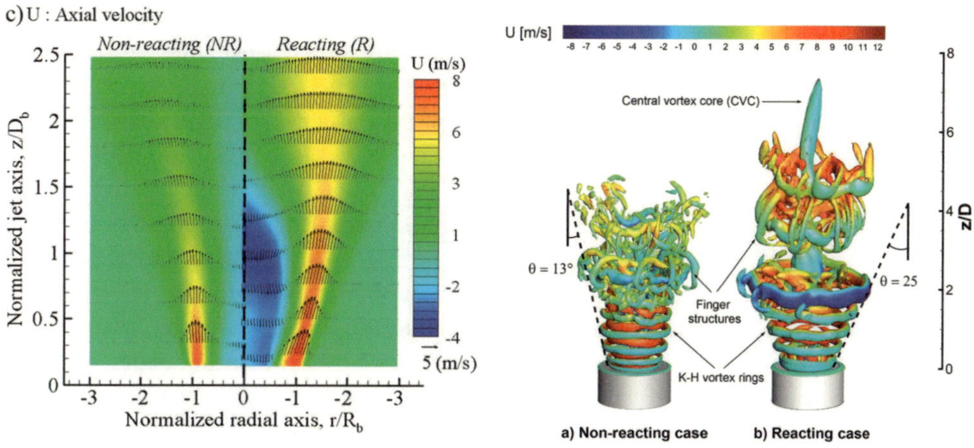

Figure 3. On the left: fields of axial velocity (U) under non-reacting and reacting conditions (S_n = 0.8 and Φ = 0.8) [13]. On the right: snapshot of iso-surfaces of Q-criterion for the non-reacting (a) and reacting (b) cases [33].

Experimental and numerical investigations have been performed on this burner configuration as shown in **Figure 3**. On the left of **Figure 3** [13], it shows contours of axial (U) velocity up to 2.5*D* downstream of the flow in non-reacting and reacting cases with the condition of a swirl number S_n = 0.8 and a global equivalence ratio Φ = 0.8. The velocity field is symmetrical about the dashed line x = 0. Note that only the left part of the field (x/R_b: from 0 to −3) is plotted in **Figure 3**. Velocity fields highlight the higher axial expansion induced by the combustion in the reacting flow. As expected, with high swirl intensity, an internal recirculation zone (IRZ) appeared for non-reacting and reacting conditions. The IRZ becomes larger by the combustion due to the change in fluid density and more intense for the reacting case compared to the non-reacting case.

On the right of **Figure 3**, the 3D coherent structures are identified using Q-criterion iso-surface for both non-reacting and reacting flows [33]. The iso-surfaces are coloured by the instantaneous axial velocity. The Q-criterion identifies vortices of an incompressible flow as connected fluid regions with a positive second invariant of the velocity gradient tensor [34]. The 3D coherent structures can reveal in a discernible manner the swirling flow instabilities. Two similar structures are present in the isothermal and the reacting flows, which are the vortex rings and the finger structures. In the near field at the burner exit, each jet is composed of two shear layers (ISL and OSL) owing to the annular geometry of the burner. More details about these works can be found in the previous paper of the authors [35–37].

5. Swirl effects on pollutant emissions

Schmittel et al. [10] showed that the use of swirl in non-premixed combustion can lead to the reduction of pollutant emissions, particularly that of nitrogen oxides. Indeed, under the swirl effect and thus the mixing improvement of reactants, the flame temperature decreases and leads to the decrease of NOx production. In addition, when the intensity of the swirl is sufficient, increasing the swirl number induces a reduction of residence time in hot areas. This also has the effect of limiting NOx formation. However, it is necessary to find a swirl intensity that achieves a compromise between on the one hand the reduction of pollutant emission and, on the other hand, the distance flame burner to prevent the flashback. The study of Coghe et al. [11] on a lean natural gas burner ($\phi = 0.69$) showed that NOx reduction can reach 30% for a swirl number between 0.7 and 0.82. Pesenti and Lybaert [38] tested in a 300 kW furnace with a swirl burner and an air co-flow by comparing three ratios of tangential airflow to total flow (air + natural gas) 0, 0.5 and 1. The authors showed that the optimum in terms of NOx emission is obtained when 50% of tangential air is injected. Burguette and Costa [39] evaluated the influence of the swirl intensity on NOx through the angle of the blades constituting the swirler. They found that at 45° the NOx rate is the most important; however, at 45° lower or higher, NOx emissions decrease. The authors explain these results by the fact that at 45° the recirculation zone starts, and therefore the residence time increases close to the burner, thus increasing the intensity of combustion and consequently the temperature and NOx formation. Cozzi and Coghe [12] studied NOx emissions on a coaxial swirled natural gas flame. They reported that the swirl intensity has a strong effect on non-premixed flames in which very low levels of NOx can be obtained at higher swirl.

Boushaki et al. [15, 40] investigated reacting swirling flows on a burner with triple jets as shown in **Figure 4**. The burner is composed of three nozzles, one central of natural gas and two laterals of pure oxygen. Each nozzle is surrounded tangentially by four small jets around the main jets. These small jets generate a swirling motion of the flow; the swirl intensity is controlled by the flow rate in the actuators. The authors reported that the swirl intensity through the flow rate of actuators affects significantly the flow field and consequently the flame behaviour (stability, shape, pollutants). For example, **Figure 4b** illustrates NOx emissions as a function of the swirl intensity controlled by the flow rate ratio *r* (ratio of flow rate of jet actuators and total flow rate). The results show that the NOx emissions with swirl intensity decrease in the first part. In this part, the NOx reductions are about 22, 36 and 60%

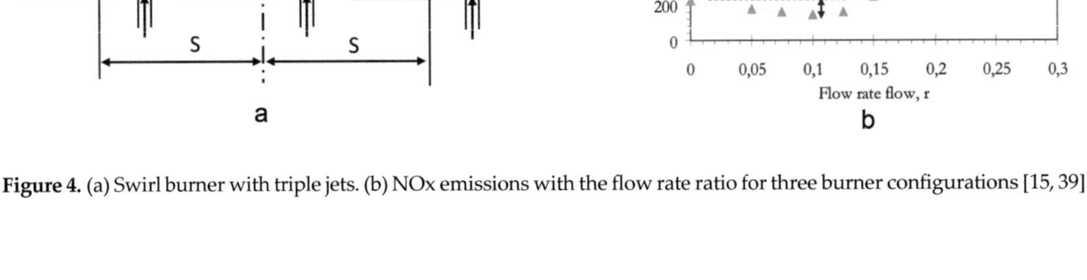

Figure 4. (a) Swirl burner with triple jets. (b) NOx emissions with the flow rate ratio for three burner configurations [15, 39].

depending to the burner configuration. From 0 to 15% of flow rate ratio, the swirling flow enhances mixing progressively and probably induces a reduction of NOx emission. Besides, the swirl by jet actuators enhances entrainment of combustion products, which decreases the flame temperature and NOx production. For the second part with higher swirl intensity, the increase of NOx production can be explained by the higher velocities of the small jet which disturb the main jets.

Boushaki et al. [13] and Nazim et al. [41] reported the effect of the swirl number on the NOx and CO emissions. They found that in the case of a swirl number of 1.4, the EICO rate is slightly lower than in the case of a swirl number of 0.8. The authors noted that the swirl intensity might tend to enhance the mixing and to increase the residence time inside the reaction zone, which promotes the CO conversion to CO_2. Concerning the NOx emissions, the authors found that increasing the swirl number tends to reduce the $EINO_x$ formation in particular for oxygen rate up to 27%. The authors explain this evolution by the decrease of flame temperature induced by the swirl through the Zeldovich mechanism. Some authors as [10, 42] by the experiments and computations demonstrated that NO_x emissions decrease when the swirl number increases, mainly because of the better mixing with combustion products in the inner recirculation zone created by high swirling flows.

Author details

Toufik Boushaki

Address all correspondence to: toufik.boushaki@cnrs-orleans.fr

ICARE CNRS, University of Orléans, France

References

[1] Beér JM, Chigier NA, editors. Combustion Aerodynamics. Florida: Krieger, Malabar; 1972

[2] Syred N, Béer JM. Combustion in swirling flows: A review. Combustion and Flame. 1974;**23**:143-201

[3] Yuasa S. Effects of swirl on the stability of jet diffusion flames. Combustion and Flame. 1986;**66**:181-192

[4] Takahashi F, Schmoll WJ. Lifting criteria of jet diffusion flames. In: 23th Symposium (International) on Combustion; Orléans, France: The Combustion Institute; 1990; pp. 375-382

[5] Cha MS et al. Effect of swirl on lifted flame characteristics in nonpremixed jets. Combustion and Flame. 1999;**117**:636-645

[6] Poireault B. Mécanismes de combustion dans un brûleur méthane-air de type swirl (40kW): Influence de l'intensité de rotation. Thèse de l'université de Potiers. 1997

[7] Feikema D, Chen RH, Driscoll JF. Enhancement of flame blowout limits by the use of swirl. Combustion and Flame. 1990;**80**:183-195

[8] Masri AR, Kalt PAM, Barlow RS. The compositional structure of swirl-stabilised turbulent nonpremixed flames. Combustion and Flame. 2004;**137**:1-37

[9] Hatami R. Reduction of the NOx emission of a closed combustion chamber by changing to air flows with swirl. Applied Energy. 1981;**9**(2):159-164

[10] Schmittel P et al. Turbulent swirling flames: Experimental investigation of the flow field and formation of nitrogen oxide. In: 28th Symposium (International) on Combustion. The Combustion Institute. 2000. pp. 303-309

[11] Coghe A, Solero G, Scribano G. Recirculation phenomena in a natural gas swirl combustor. Experimental Thermal and Fluid Science. 2004;**28**:709-714

[12] Cozzi F, Coghe A. Effect of air staging on a coaxial swirled natural gas flame. Experimental Thermal and Fluid Science. 2012;**43**:32-39

[13] Boushaki T, Merlo N, Chauveau C, Gökalp I. Study of pollutant emissions and dynamics of non-premixed turbulent oxygen enriched flames from a swirl burner. Proceedings of the Combustion Institute. 2017;**36**:3959-3968

[14] Iyogun CO, Birouk M, Kozinski JA. Experimental investigation of the effect of fuel nozzle geometry on the stability of a swirling non-premixed methane flame. Fuel. 2011;**90**(4):1416-1423

[15] Boushaki T, Sautet JC, Labegorre B. Control of flames by radial jet actuators in oxy-fuel burners. Combustion and Flame. 2009;**156**:2043-2055

[16] Elbaz AM, Robert WL. Investigation of the effects of quarl and initial conditions on swirling non-premixed methane flames: Flow field, temperature, and species distributions. Fuel. 2016;**169**:120-134

[17] Sheen HJ, Chen WJ, Jeng SY, Huang TL. Correlation of swirl number for a radial-type swirl generator. Experimental Thermal and Fluid Science. 1996;**12**:444-451

[18] Gupta AK, Lilley DG, Syred N. Swirl Flows. United Kingdom: Abacus Press; 1984

[19] Ahmed SF, Balachandran R, Marchione T, Mastorakos E. Spark ignition of turbulent nonpremixed bluff-body flames. Combustion and Flame. 2007;**151**:366-385

[20] Mansouri Z, Aouissi M, Boushaki T. Detached eddy simulation of high turbulent swirling reacting flow in a premixed model burner. Combustion Science and Technology. 2016;**188**(11-12):1777-1798

[21] Galley D, Ducruix S, Lacas F, Veynante D. Mixing and stabilization study of a partially premixed swirling flame using laser induced fluorescence. Combustion and Flame. 2011;**158**:155-171

[22] Stöhr M, Arndt CM, Meier W. Transient effects of fuel-air mixing in a partially-premixed turbulent swirl flame. Proceedings of the Combustion Institute. 2015;**35**(3):3327-3335

[23] Steinberg AM, Arndt CM, Meier W. Parametric study of vortex structures and their dynamics in swirl-stabilized combustion. Proceedings of the Combustion Institute. 2013;**34**:3117-3125

[24] Mansouri Z, Aouissi M, Boushaki T. Numerical computations of premixed propane flame in a swirl stabilized burner: Effects of hydrogen enrichment swirl number and equivalence ratio on flame characteristics. International Journal of Hydrogen Energy. 2016;**41**:9664-9678

[25] Huang RF, Duc LM, Hsu CM. Flow and mixing characteristics of swirling double-concentric jets influenced by a control disc at low central jet Reynolds numbers. International Journal of Heat and Fluid Flow. 2016;**00**:1-14

[26] Jourdaine P, Mirat C, Caudal J, Lo A, Schuller T. A comparison between the stabilization of premixed swirling CO_2-diluted methane oxy-flames and methane/air flames. Fuel. 2016

[27] Benim AC, Iqbal S, Meier W, Joos F, Wiedermann A. Numerical investigation of turbulent swirling flames with validation in a gas turbine model combustor. Applied Thermal Engineering. 2017;**110**:202-212

[28] Khalil Ahmed EE, Gupta KA. Towards distributed combustion for ultra low emission using swirling and non-swirling flowfields. Applied Energy. 2014;**121**:132-139

[29] Orbay RC, Nogenmyr KJ, Klingmann J, Bai XS. Swirling turbulent flows in a combustion chamber with and without heat release. Fuel. 2013;**104**:133-146

[30] Candel S, Durox D, Schuller T, Palies P, Bourgouin JF, Moeck J, et al. Progress and challenges in swirling flame dynamics. Comptes Rendus Mecanique. 2012;**340**:758-768

[31] Von Lavante E, Yao J. Numerical investigation of turbulent swirling flows in axisymmetric internal flow configurations. Flow Measurement and Instrumentation. 2012;**25**:63-68

[32] Cozzi F, Coghe A, Sharma R. Analysis of local entrainment rate in the initial region of isothermal free swirling jets by Stereo PIV. Experimental Thermal and Fluid Science. 2018;**94**:281-294

[33] Mansouri Z, Boushaki T. Experimental and numerical investigation of turbulent isothermal and reacting flows in a non-premixed swirl burner. International Journal of Heat and Fluid Flow. 2018;**72**:200-213

[34] Hunt JCR, Wray AA, Moin P. Eddies, streams, and convergence zones in turbulent flows. Report CTR-S88; Stanford: Stanford University, Center for Turbulence Research; 1988. pp. 193-208

[35] Merlo N, Boushaki T, Chauveau C, de Persis S, Pillier L, Sarh B, et al. Combustion characteristics of methane-oxygen enhanced air turbulent non-premixed swirling flames. Experimental Thermal and Fluid Science. 2014;**56**:53-60

[36] Boushaki T, Koched A, Mansouri Z, Lespinasse F. Volumetric velocity measurements (V3V) on turbulent swirling flows. Flow Measurement and Instrumentation. 2017;**54**:46-55

[37] Zaidaoui H, Boushaki T, Sautet JC, Chauveau C, Sarh B, Gökalp I. Effects of CO_2 dilution and O_2 enrichment on non-premixed turbulent CH_4-air flames in a swirl burner. Combustion Science and Technology. 2018;**190**(5):784-802

[38] Pesenti B, Lybaert P. Numerical and experimental investigation of a 300 kW swirl gas burner: Flame shape and NOx emissions. In: 2nd European Combustion Meeting; Louvain-la-Neuve, Belgium; 2005

[39] Burguette M, Costa M. NOx emissions from unconfined swirl flames. In: 7th European Conference on Industrial Furnaces and Boilers INFUB7, Porto. 2006

[40] Boushaki T, Sautet JC. PIV measurements on oxy-fuel burners. In: The Particle Image Velocimetry: Characteristics, Limits and Possible Applications. InTech; 2012. pp. 123-148. ISBN 978-953-51-0625-8

[41] Merlo N, Boushaki T, Chauveau C, de Persis S, Pillier L, Sarh B, et al. Experimental study of oxygen enrichment effects on turbulent nonpremixed swirling flames. Energy & Fuels. 2013;**27**:6191-6197

[42] Frassoldati A, Frigerio S, Colombo E, Inzoli F, Faravelli T. Chemical Engineering Science. 2005;**60**(11):2851-2869

Chapter 2

Coherent Structures in the Near Field of Swirling Turbulent Jets and Flames Investigated by PIV and PLIF

Vladimir Dulin, Aleksei Lobasov,
Dmitriy Markovich and Sergey Alekseenko

Additional information is available at the end of the chapter

http://dx.doi.org/10.5772/intechopen.79896

Abstract

The results of the experimental study of coherent flow structures in turbulent jets with different swirl rates are reported. The focus is placed on analysis of their impact on mixing of a passive scalar in the jets and on regular flame front deformations during combustion of fuel-lean and fuel-rich methane/air mixtures in case of a strongly swirling jet with vortex breakdown. The measurements are performed by applying simultaneously the particle image velocimetry (PIV) and planar laser-induced fluorescence (PLIF) techniques (acetone and HCHO for the nonreacting and reacting flows, respectively). The PIV data are processed by a proper orthogonal decomposition (POD), and the PLIF data are conditionally sampled according to the correlation coefficients of the velocity POD modes. The coherent velocity fluctuations are associated with regular patterns in the concentration fluctuation fields and flame front deformations. These patterns correspond to unsteady mixing by large-scale vortex structures in the outer mixing layer and also to variations of the entrainment rate for the flows with swirl.

Keywords: swirling jet, mixing, swirling flame, vortex breakdown, coherent structures, precessing vortex core

1. Introduction

Swirl is often superimposed to jet flows to promote mixing and heat and mass transfer in the initial region of the flow [1–4]. Whereas ring-like vortices are formed in the shear layer of the nonswirling and weakly swirling jets, helical vortex structures are formed in the mixing layer of strongly swirling jets and considered to enhance mixing [4–8]. When the swirl rate exceeds a certain critical value, a breakdown of the vortex occurs in swirling jets [6, 7, 9–12]. The vortex breakdown corresponds to formation of a wake region or recirculation zone at the jet axis and intensifies mixing even more via unsteady flow dynamics. The low velocity region at the low-pressure core of the swirling flow provides favorable local conditions for successful ignition and stable combustion of flames [1, 13]. Thus, swirling jet flows are often implemented in combustion chambers and furnaces.

Structure of swirling flows with combustion has been studied extensively in a number of experiments. Heat release and local density decrease during combustion strongly affect the shape of the recirculation zone [14, 15]. Hot combustion products tend to concentrate at the vortex core and are captured by the central recirculation zone [16–20]. Moreover, combustion is known to induce vortex breakdown, which was absent for the nonreacting flow before ignition of the flame [21–23]. Besides, amplitude of the vortex core precession is found to be suppressed by combustion in some studies, whereas in the other, it was not strongly affected [24, 25]. Detailed experimental data on flow structure of swirling flows with combustion is necessary for verification of numerical simulations, which are currently performed with eddy-resolving methods [26–31].

Presently, the impact of unsteady swirling flow dynamics, such as precession of the vortex core, which is found to produce intensive velocity and pressure pulsations in jets with vortex breakdown, on unsteady combustion is not completely understood [32]. In particular, the impact of the precessing vortex core on flame instabilities and unsteady operation of swirl-stabilized combustors is still a debated issue [33]. Ruith et al. [11] and Liang and Maxworthy [7] suggest that formation of the recirculation zone triggers a global instability mode, which results in unsteady dynamics of swirling jets, related to intensive velocity and pressure fluctuations during precession of the swirling vortex core near the nozzle. Oberleithner et al. [12] have suggested that during the increase of the jet swirl rate the central recirculation zone first appears in an intermittent manner and then remains permanently. Further increase of the swirl rate gives rise to a global flow instability to a helical mode, corresponding to precession of the swirling jet.

From conditionally sampled velocity measurements by the laser Doppler velocimetry technique, Cala et al. [34] have detected a coherent structure in a high-swirl jet, consisted of a spiraling vortex core and secondary helical large-scale vortex structures in the inner and outer mixing layers. Later, similar structures were detected from 2D particle image velocimetry (PIV) measurements for high-swirl nonreacting jets [35, 36] and flames [37–39]. Later, the presence of double-helical coherent vortex structure was confirmed from direct 3D PIV measurements in [8, 40]. However, there is still a lack of experimental studies on mixing in swirling turbulent jets and quantitative analysis of large-scale vortices contribution to this process.

Dynamics of flow and flame in swirl combustors is studied successfully by combination of PIV and planar laser-induced fluorescence (PLIF) of OH* (see [41, 42]). The hydroxyl radical OH*, produced in flame front and present in hot combustion products, is commonly used for tracking of the front. Stöhr et al. [42] have found that large-scale vortices improve mixing between the combustion products and reactants. Meanwhile, Boxx et al. [41] have observed events of local flame extinction during its interaction with large-scale vortices in the high-swirl flow. Nevertheless, further detailed analysis of the reaction zone shape correlations with large-scale flow motions in swirling jet flows is desirable.

Formaldehyde (HCHO) is an important combustion intermediate, occurring in lower-temperature regions of hydrocarbon-fueled flames. It plays an important role in several combustion processes, including fuel oxidation and autoignition. It appears in the initial step of the HCHO → HCO → CO oxidation pathway of conventional hydrocarbons [43]. High concentration of HCHO specifies preheat zone of hydrocarbon flames. One of the most prevalent strategies for HCHO PLIF measurements is the excitation of $4^1{}_0$ transition by using the third harmonic of Nd:YAG laser radiation at 355 nm [44].

The aim of the current study is to analyze the impact of large-scale vortex structures on mixing in turbulent jets with different swirl rates on the basis of the combined PIV and PLIF measurements and to reveal regular flame front deformations during combustion of fuel-lean and fuel-rich flames in the swirling jet with high swirl rate and vortex breakdown.

2. Experimental setup and data processing

Measurements were carried out for swirling jets and premixed methane/air flames at the atmospheric pressure. The flow was organized in an open combustion rig by using a contraction axisymmetric nozzle with a changeable vane swirler installed inside to generate jet flows with swirl. The outlet diameter of the nozzle was d = 15 mm (see [37] for the details). The swirl rate S was estimated based on the geometric parameters of the swirler [1]:

$$S = \frac{2}{3}\left(\frac{1-(d_1/d_2)^3}{1-(d_1/d_2)^2}\right)\tan(\psi) \quad (1)$$

where d_1 = 7 mm is the diameter of the center body supporting the vanes, d_2 = 27 mm is the external diameter of the swirler, and ψ is the vanes inclination angle relative to the axis. Three cases of the swirl rate are considered, viz., nonswirling jet without swirler (S = 0) and swirling jets for ψ = 30, 45, and 55°, corresponding to S = 0.41, 0.7, and 1.0, respectively, whereas the critical value for the vortex breakdown in jet flows is $S \approx 0.6$ [14, 45–47].

The air flow was supplied from a pressure line, and its flow rate was precisely controlled by mass flow meters (Bronkhorst). To introduce acetone vapor into the nonreacting jet flows, a part of the air flow was bubbled through liquid acetone, contained in a thermostabilized container at a fixed temperature. The concentration of the acetone in the jet was below 3%. The Reynolds number based on the flowrate of air was fixed as 5000 (bulk velocity for the air jet was U_0 = 5.0 m/s).

For the reacting flows, the container with acetone was not used. Instead, the air was premixed with methane, supplied by another set of Bronkhorst mass flow meters. The equivalence ratio φ of the methane/air mixtures issued from the nozzle was equal to 0.7 and 2.5. To provide PIV measurements, the jet flow was seeded by 0.5 μm TiO_2 particles. The surrounding air was seeded by using a fog generator.

A sketch of the PIV/PLIF experimental setup is shown in **Figure 1**. Two CCD PIV cameras (ImperX IGV-B2020) were oriented horizontally as shown in **Figure 1**. The cameras were

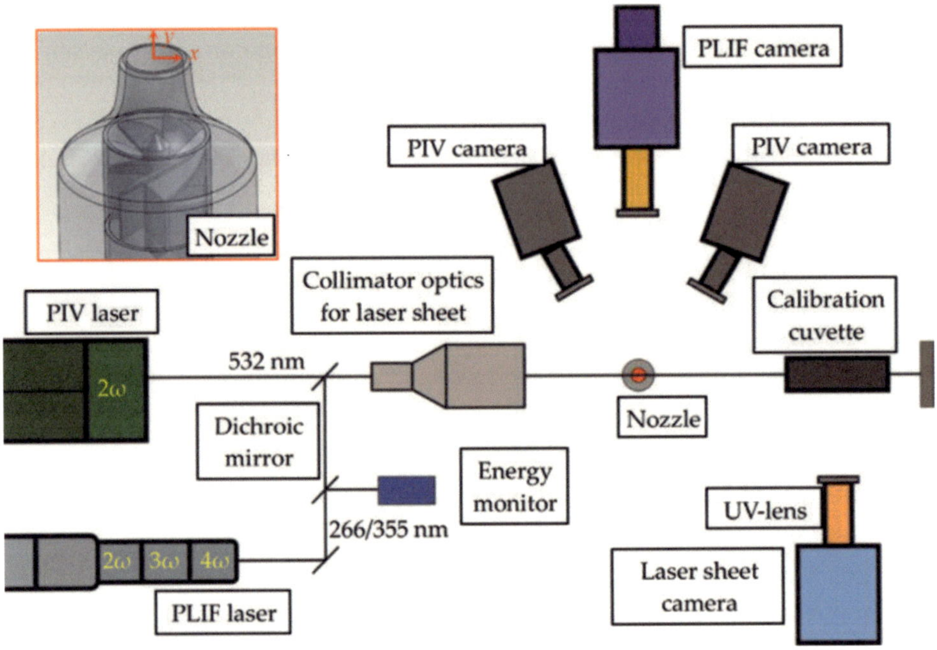

Figure 1. Sketch of the PIV/PLIF setup and geometry of the nozzle (inset).

equipped with narrow band-pass optical filters (532 ± 10 nm). A double-head Nd:YAG laser (Quantel EverGreen) was used to illuminate the tracer particles in the central plane of the flow for the PIV measurements. Duration of each laser pulse was approximately 10 ns. The time interval between two PIV laser pulses for the reacting flows was 35 μs.

The velocity fields were evaluated by an adaptive iterative cross correlation algorithm with continuous image shift and deformation (see [48] for the definition), included an in-house software "ActualFlow." The final size of the interrogation areas was 32 × 32 pixels. The spatial overlap rate between the neighbor interrogation areas was 50%. Calibration of the stereo PIV and PLIF cameras was performed by using a plane calibration target and third-order polynomial transform. This was done by processing images of the calibration target placed in five different positions in the normal-to-plane direction with the step of 0.5 mm. In addition, to minimize the calibration error, an iterative correction procedure of possible misalignment between the laser sheet and the target plane was applied [49]. For almost the entire measurement domain, the mismatch between the actual marker locations on the target and their coordinates in the obtained calibration model was below 1 pixel.

The fourth harmonic (266 nm) of a pulsed Nd:YAG laser (Quantel Brilliant B) was used for excitation of the acetone fluorescence. During excitation of the HCHO fluorescence (4^1_0 transition of A–X band) for the reacting flows, the third harmonic (355 nm) of the same laser was used. Duration of each laser pulse was approximately 10 ns. The laser beams for the PLIF and PIV systems were combined by using a dichroic mirror and converted into a collimated laser sheet with the width of 50 mm. The laser sheet thickness was below 0.8 mm in the measurement region.

Fluorescence of HCHO and acetone was collected by a system of UV-sensitive image intensifier (LaVision IRO, photocathode S20 multialkali provided quantum efficiency about 25% for wavelengths in UV spectral region) and sCMOS camera (LaVision Imager sCMOS, 16 bit images with resolution of 2560 × 2160 pixels), equipped with a UV lens (100 mm, f# = 2.8) and band-pass optical filter. The exposure time for each PLIF image was 200 ns. The PLIF signal was collected almost in the center of the time interval between two pairs of the PIV laser pulses.

The raw PLIF images contained different types of systematic and random errors. The systematic errors were produced by spatially nonuniform laser sheet intensity, nonuniform spatial sensitivity of the photocathode and CCD, dark current of the sensors, and background signal (due to possible reflections, etc.). A set of post-processing algorithms were applied to the PLIF images to remove the dark current and background intensity and to correct for the spatial nonuniformity of the laser sheet intensity and sensitivity of the sensors. An additional ICCD camera (PCO DicamPro) was used to monitor the spatial distribution of energy in each pulse of the PLIF lasers by using a quartz cuvette filled with uniform solution of a fluorescent dye (Rhodamine 6G).

Typical magnification of the PIV and PLIF systems was 29.9 and 35.0 pixels per mm, respectively. The optical resolution of the PLIF system corresponded to a Gaussian-type smoothing with the full width at the half maximum of approximately 15 pixels. The resolution was evaluated based on a "knife-edge technique" similar to that in [50]. To fit grids of the PIV and PLIF data, the latter was spatially averaged over domains with the size of 0.96 mm and 50% overlap rate. The laser sheet thickness in the region of interest for both systems was approximately 0.8 mm.

To reveal coherent structures in the flows, the fluctuating velocity data set $W = [\mathbf{u}'(\mathbf{x}, t_1)\ldots\mathbf{u}'(\mathbf{x}, t_N)]$ for each flow case was processed by a snapshot proper orthogonal decomposition (POD) [51]. POD is based on a singular value decomposition:

$$\mathbf{u}'(\mathbf{x}, t_k) = \sum_{q=1}^{N} \alpha_q(t_k)\sigma_q\boldsymbol{\varphi}_q(\mathbf{x}) \text{ or in a matrix form } W = U\Sigma V^H, \tag{2}$$

$$\text{where } \int_\Omega \boldsymbol{\varphi}_i\boldsymbol{\varphi}_j d\mathbf{x} = \delta_{ij} \text{ and } \frac{1}{N}\sum_{k=1}^{N} \alpha_i(t_k)\alpha_j(t_k) = \delta_{ij} \tag{3}$$

Each set of the fluctuating velocity fields $\mathbf{u}'(\mathbf{x}, t_k)$ is represented as a finite series of the products of the spatial orthonormal basis functions $\boldsymbol{\phi}_q$ ($U = [\boldsymbol{\phi}_1\ldots\boldsymbol{\phi}_N]$) with nondimensional temporal coefficients α_q (corresponding to the conjugate transpose of a matrix V that is composed of the right-singular vectors of W) and singular values σ_q ($\Sigma = \mathrm{diag}[\sigma_1\ldots\ \sigma_N]$). N is the number of the snapshots in the set. The singular values characterize the amplitude (square root of the kinetic energy of the velocity fluctuations) of each POD mode in the data sequence and are equal to square root of the eigenvalues of the covariance matrix WW^H. For the used snapshot POD method, the POD modes can be represented as:

$$\boldsymbol{\varphi}_q(\mathbf{x}) = \sum_{k=1}^{N} a_q^k \mathbf{u}'(\mathbf{x}, t_k)\sigma_k^{-1} \text{ or } U = W\Sigma^{-1}V \tag{4}$$

To reveal coherent structure effect on mixing and flame front deformation, fluctuations $I'(\mathbf{x}, t)$ of the PLIF data are conditionally sampled according to the temporal coefficients $a^k{}_q$ (from matrix V) of the POD modes:

$$c_q(\mathbf{x}) = \frac{u_{rms}}{I_{rms}} \sum_{k=1}^{N} a_q^k I'(\mathbf{x}, t_k)\sigma_k^{-1}, \text{ where } u_{rms}^2 = \int_\Omega \mathbf{u}'^2 d\mathbf{x} \text{ and } I_{rms}^2 = \int_\Omega I'^2 d\mathbf{x} \tag{5}$$

3. Results

3.1. Swirling jets without combustion

Figure 2 shows the time-averaged velocity fields and turbulent kinetic energy for the nonreacting jets. The core of the nonswirling jet (S = 0) is surrounded by the circular mixing layer, where velocity fluctuations grow downstream. For the case S = 0.41 (referred to as low-swirl jet), a wake region is formed at the jet axis, where the average axial velocity remains positive. The velocity fluctuations for the outer mixing layer of the low swirl are considerably greater. Also, their intensity at the onset of the central wake region reached 35% of U_0. For the cases S = 0.7 and 1.0 (referred to as high-swirl jets), the black solid line surrounds the region with negative values of the mean axial velocity, corresponding to the central recirculation zone.

Thus, there are two mixing layers in the flow of the swirling jets, viz., the inner mixing layer between the annular swirling jet and the central recirculation zone/wake region and the outer mixing layer between the jet and the surrounding air. The swirl results in a greater opening angle for the jet. It is increased from approximately 11 to 37° (half angle) for the considered range of swirl rates.

The normalized spatial distributions of the time-averaged acetone concentration are shown in **Figure 3**. The swirl results in a faster mixing of the jet with the surrounding air. The contour lines show values of the local variance of the concentration. Thus, the concentration fluctuates in the outer mixing layer. The swirl decreases the length of the mixing region, viz., it is less than one nozzle diameter for the case S = 1.0. Thus, the flow swirl and vortex breakdown dramatically promote the mixing.

Examples of the instantaneous velocity and concentration fields, measured simultaneously and plotted in **Figure 4**, show an effect of large-scale vertical structures on turbulent mixing in the studied jets. The large-scale vortex structures are visualized by regions with positive values of a 2D modification (Eq. (6)) of Q-criterion [52], viz., $Q^{2D} > 5U_0^2 d^{-2}$. For the nonswirling jet, it is expected that the vortex cores in the outer mixing layer correspond to ring-like vortices. They engulf and entrain the surrounding air, whereas the jet core remains almost unmixed. For the low-swirl jet, the acetone spreads faster into the surrounding air and the mixing is more efficient.

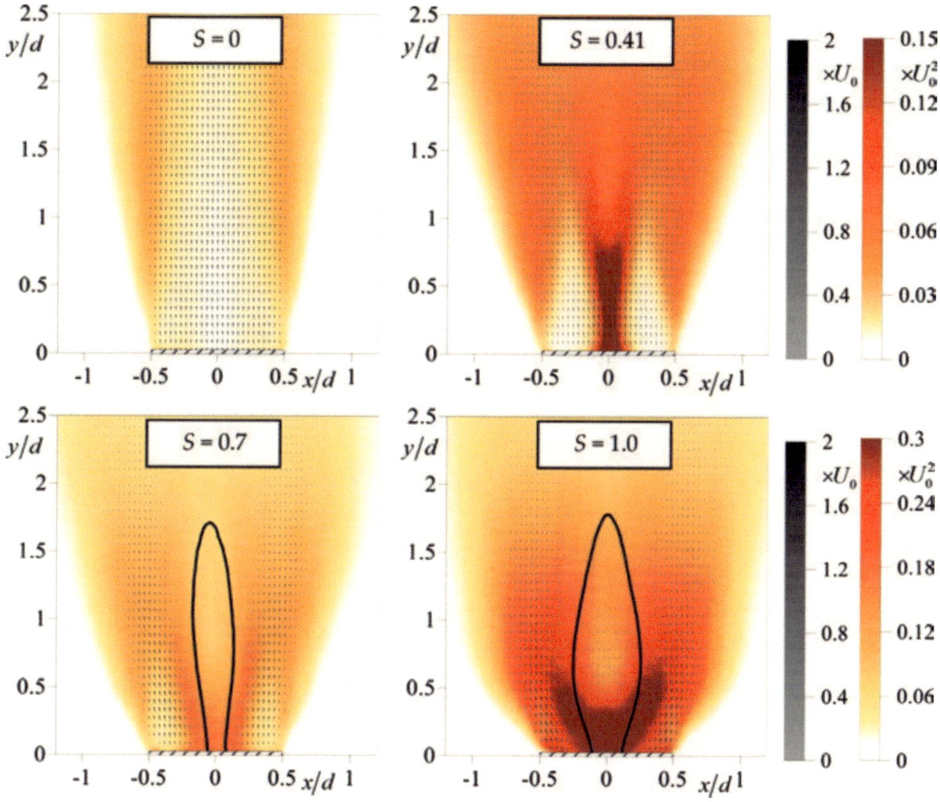

Figure 2. Time-averaged velocity fields and distributions of the turbulent kinetic energy for nonreacting swirling jets. Black solid line includes regions with negative axial velocity.

$$Q^{2D} = -\frac{1}{2}\frac{\partial u_x}{\partial x}\frac{\partial u_x}{\partial x} - \frac{1}{2}\frac{\partial u_y}{\partial y}\frac{\partial u_y}{\partial y} - \frac{\partial u_x}{\partial y}\frac{\partial u_y}{\partial x}. \quad (6)$$

Whereas the vortex structures present both in the inner and outer mixing layers of the low-swirl jet, their contribution to the mixing near the wake region is small. During the mixing in the high-swirl jets, some amount of the surrounding air is captured by the central recirculation zone and mixed with annular jet, issued from the nozzle. Thus, the contribution of large-scale vortex structures to the jet dilution in the inner mixing layer increases with swirl rate.

To reveal coherent structures in the flows and to analyze their contribution to mixing, POD is applied to the sets of the measured velocity fields. The POD spectra of the velocity fluctuations in the nonreacting jets for different swirl rates are shown in **Figure 5**. For the high-swirl jets with vortex breakdown, there are two most energetic modes. They contain approximately 9 and 16% of the spatial-averaged turbulent kinetic energy for the cases of S = 0.7 and 1.0, respectively, whereas the energy of rest modes is below 2%. Values of the temporal coefficients for the first two modes for S = 1.0, shown in **Figure 5**, are scattered around a circle-like figure, indicating that these modes are statistically correlated.

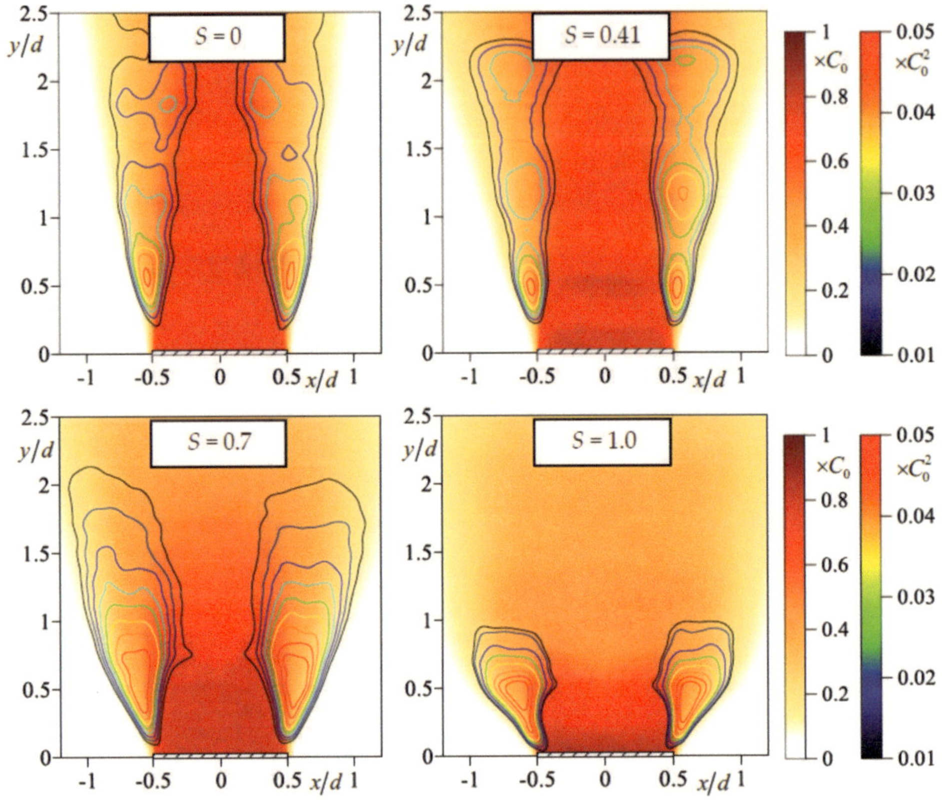

Figure 3. Time-averaged values and variance of the passive scalar concentration for nonreacting turbulent jets with different swirl rate.

Spatial distributions of the first four POD modes for the nonswirling jet with corresponding conditionally sampled fluctuations of the acetone concentration are shown in **Figure 6**. The distributions of the fluctuations correspond to growing downstream vortex structures and expanding traveling waves of the concentration fluctuations. The first and second POD modes appear to be shifted by $\pi/2$ phase and are expected to be related to growing downstream ring-like vortices. The third and fourth modes are likely to be associated with pairing process of these ring-like vortices at approximately 1.5d downstream the nozzle exit and related modulation of amplitude of the subsequent vortices upstream via pressure feedback mechanism [53].

For the low-swirl jet (see **Figure 7**), the first POD mode is related to oscillations of the axial velocity near the central wake, which is in agreement with the scenario of the vortex breakdown by Oberleithner et al. [12], where formation of the permanent recirculation zone is preceded by an intermittent formations of the reverse flow at the jet axis. It is found that such fluctuations are coherent with large-scale variation of the entrainment rate downstream. The second POD mode is presumably related to precession of the vortex core upstream the wake region and also correlated with downstream variation concentration fluctuations. The third and fourth POD modes correspond to traveling waves of the concentration fluctuations in the outer mixing layer, produced by growing downstream large-scale toroidal vortex structures.

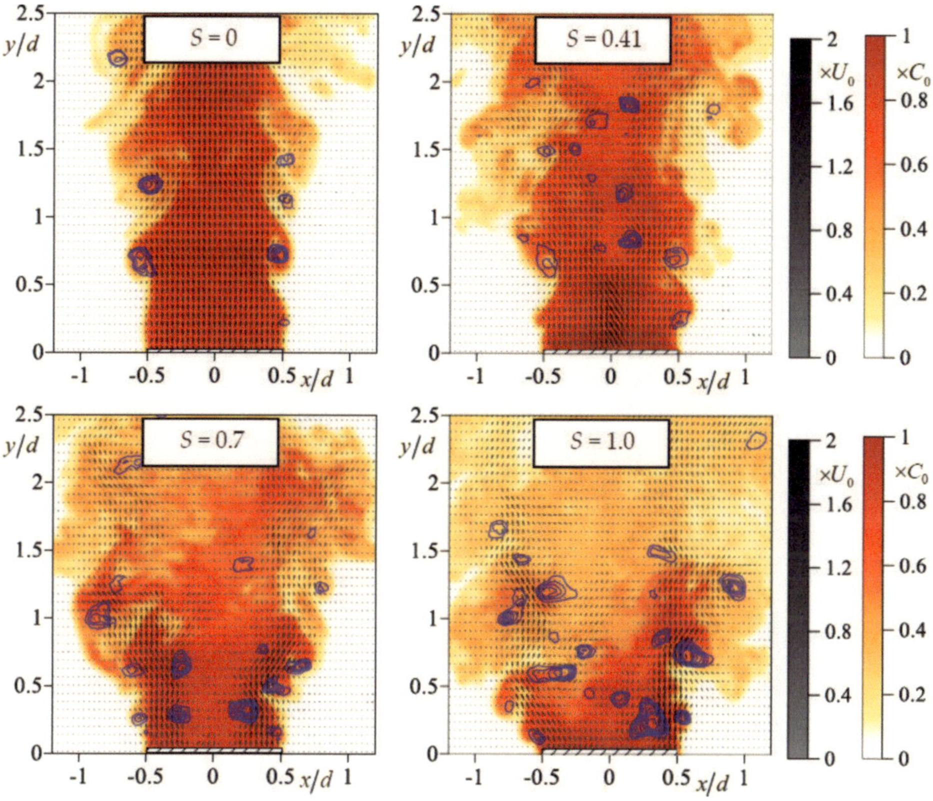

Figure 4. The instantaneous velocity and concentration snapshots for turbulent jets with different swirl rate. Large-scale vortical structures are visualized by regions with positive Q-criterion ($Q^{2D} \geq 5U_0^2d^{-2}$ with the step of $2.5U_0^2d^{-2}$).

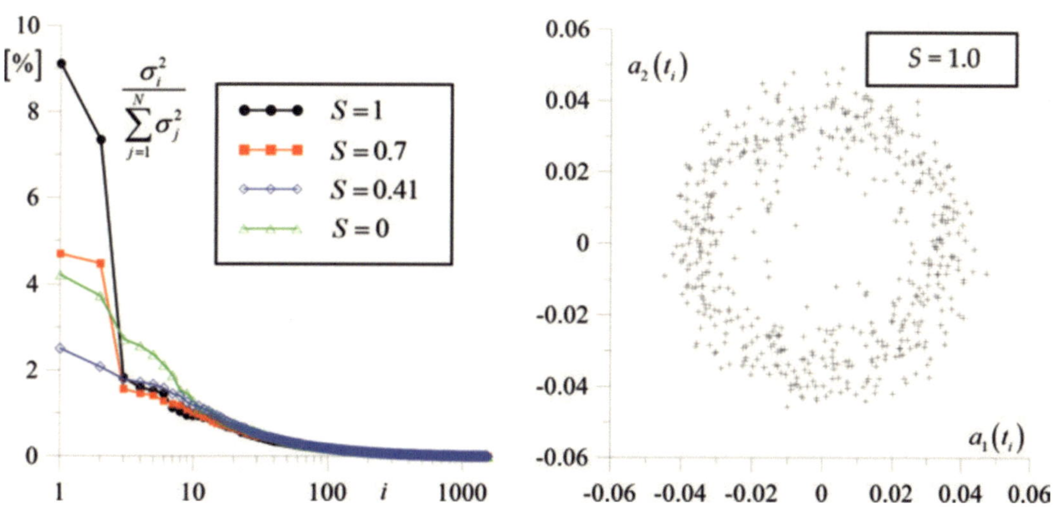

Figure 5. POD spectra for nonreacting jets and temporal coefficients for the first two POD modes for the high-swirl jet.

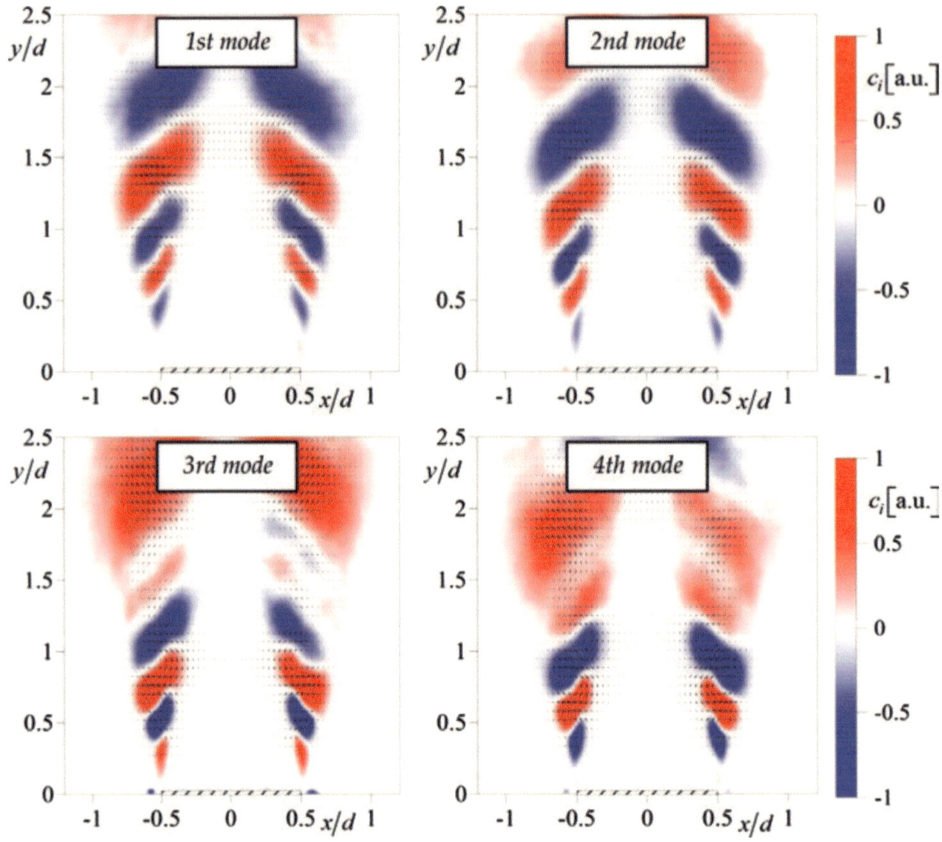

Figure 6. First four POD modes of the velocity fluctuations and corresponding conditionally sampled fluctuations of the concentration for a nonswirling jet ($S = 0$).

Figures 8 and **9** show the first four POD modes for the high-swirl jets for $S = 0.7$ and 1.0, respectively. The first two POD modes correspond to transverse velocity fluctuations around the nozzle exit, correlated with the velocity fluctuations in the inner and outer mixing layer. The concentration fluctuations related to these modes correspond to expanding downstream traveling waves. In analogy to the nonswirling jet, the first and second POD modes appear to be shifted by the phase of $\pi/2$. The main difference between the cases $S = 0.7$ and 1.0 is that in the latter case the coherent fluctuations of the concentration take place both in the inner and outer mixing layers.

The results for the high-swirl jets are in agreement with the conclusions of the previous 2D PIV study in [37], where it was proposed that these two POD modes correspond to two orthogonal cross planes of a single rotating coherent structure, consisting of a precessing spiraling vortex core (in the inner mixing layer) and a secondary helical vortex structure (in the outer mixing layer). This assumption was later supported in studies of the nonreacting swirling jets by 3D PIV [8]. Thus, the outer helical vortex of the coherent structure produces regular large-scale structures in the spatial distributions of the concentration fluctuations.

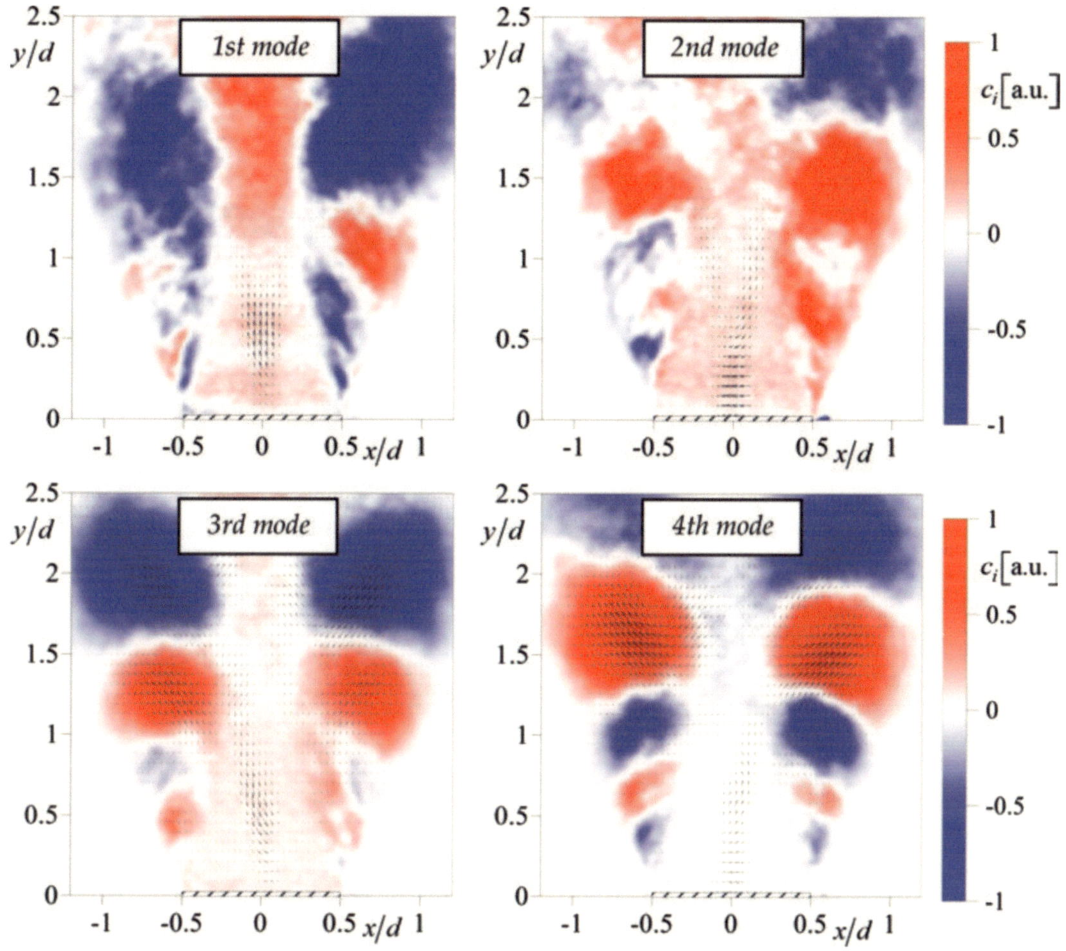

Figure 7. First four POD modes of the velocity fluctuations and corresponding conditionally sampled fluctuations of the concentration for a low swirling jet ($S = 0.41$).

3.2. High-swirl jets with combustion

Figure 10 shows the photograph, time-averaged velocity field, and HCHO PLIF signal for the high swirling jet with combustion. Two flame cases are considered, viz., combustion of the fuel-lean and fuel-rich mixtures with the equivalence ratio of $\varphi = 0.7$ and 2.5, respectively. The black solid lines in the time-averaged velocity fields surround regions with negative values of the mean axial velocity, corresponding to the central recirculation zone. Note that the shape of the recirculation zone for these two cases is very similar. In general, there are two mixing layers in the flows as it is the case for the nonreacting high-swirl jet.

For the fuel-lean flame ($\varphi = 0.7$), the annular jet flow envelopes the central recirculation zone, containing a weak reverse flow. According to the PLIF signal, the flame front is located in the inner mixing layer of the jet flow. On average, the combustion of the fuel-rich mixture takes

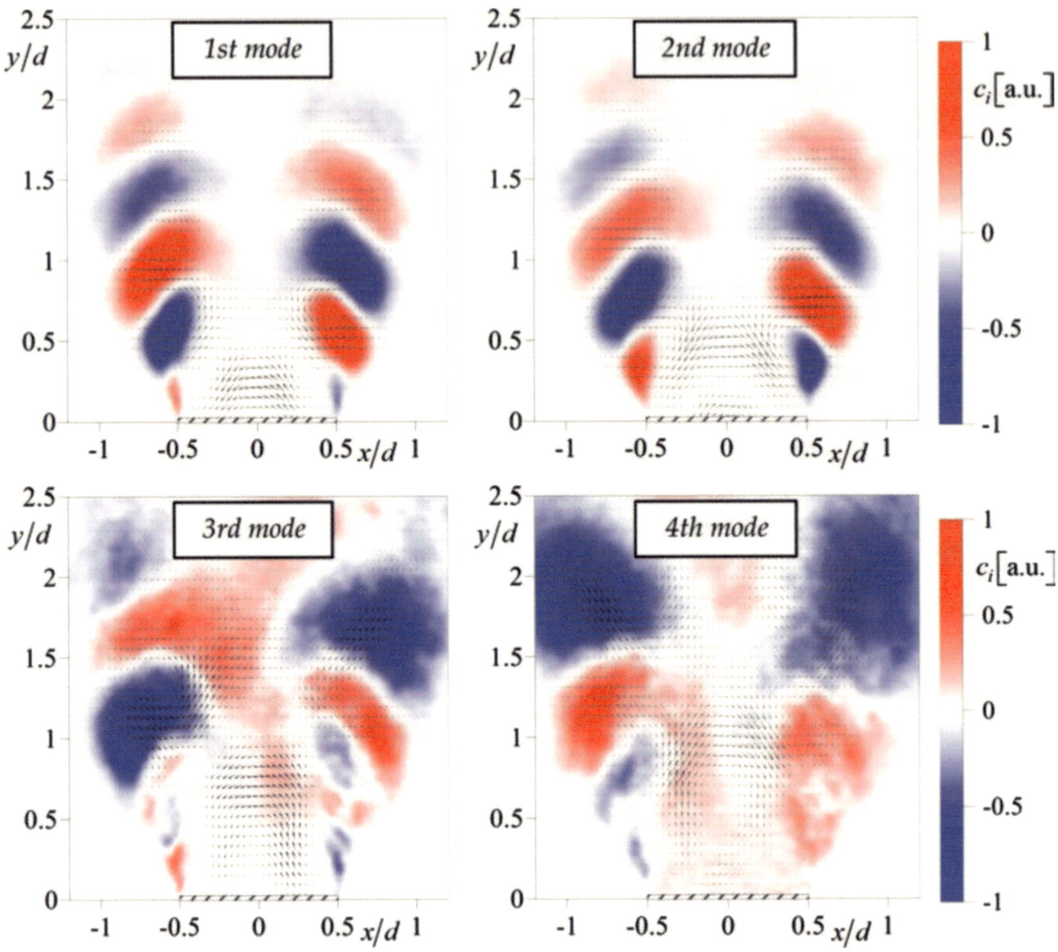

Figure 8. First four POD modes of the velocity fluctuations and corresponding conditionally sampled fluctuations of the concentration for a high swirling jet ($S = 0.7$).

place both in the outer (above $0.7d$ from the nozzle exit) and in the inner (around the central recirculation zone) mixing layers.

Figure 11 shows the instantaneous snapshots of the velocity and HCHO PLIF signal for the high-swirl jets with combustion. Large-scale vortex structures are visualized by the 2D modification of Q-criterion. For the fuel-lean mixture, there are pairs of large-scale vortices, which corrugate the flame front (indicated by arrows). The vortices in the pair envelope the flame front from opposite sides (one vortex is located in the inner mixing layer, whereas another one is in the outer mixing layer). In the previous study of [39], it was suggested that these vortex pairs are cross sections of two helical vortex structures with one present in the inner mixing layer and another one located in the outer mixing layer.

The instantaneous velocity and HCHO PLIF data for the fuel-rich flame illustrate that the flame front surrounds the central reticulation zone with a complex shape, corresponding to downward flow between the large-scale vortex structures in the inner mixing layer. The flame

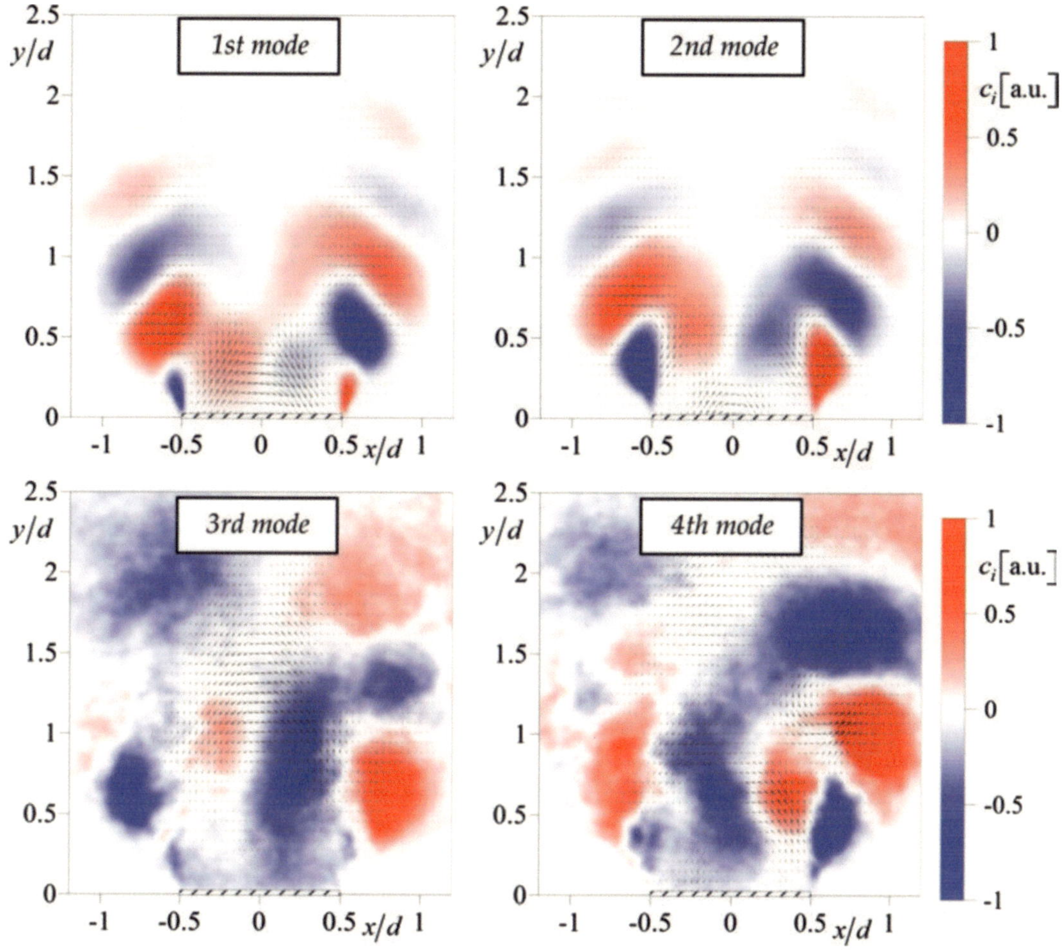

Figure 9. First four POD modes of the velocity fluctuations and corresponding conditionally sampled fluctuations of the concentration for a high swirling jet ($S = 1.0$).

front is also located in the inner mixing layer, where turbulent mixing of the combustion products with the surrounding air is followed by their afterburning. In general, the current PIV/PLIF data supports results of the previous study in [37], where it was concluded that the large-scale helical vortex structure in the outer mixing layer promotes stabilization of the fuel-rich flame via enhanced turbulent mixing.

To reveal coherent structure effect on the flame front shape, the fluctuating velocity data sets are processed by the POD. The POD spectra for the high-swirl flows without combustion and with combustion of the fuel-lean and fuel-rich mixtures are shown in **Figure 12**. For the flows with combustion, amplitude (i.e., the spatial-averaged variance of the velocity fluctuations; note that the density is not constant) of the first two POD modes is considerably smaller in comparison to that for the nonreacting flow. The spatial distributions of the first four POD modes for the reacting flows with $\varphi = 0.7$ and 2.5 are shown in **Figures 13** and **14**, respectively.

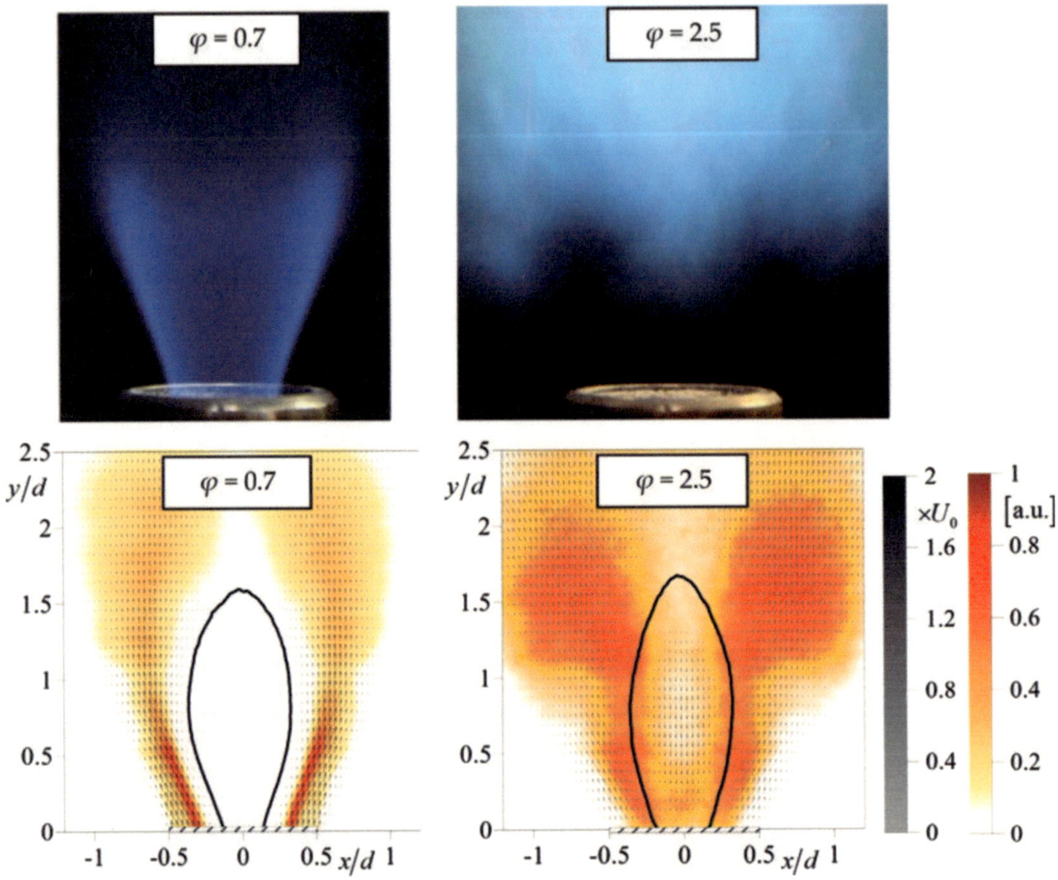

Figure 10. Photographs and time-averaged velocity field and HCHO PLIF for the high-swirl jets ($S = 1.0$) with combustion. Black solid line includes region with negative axial velocity.

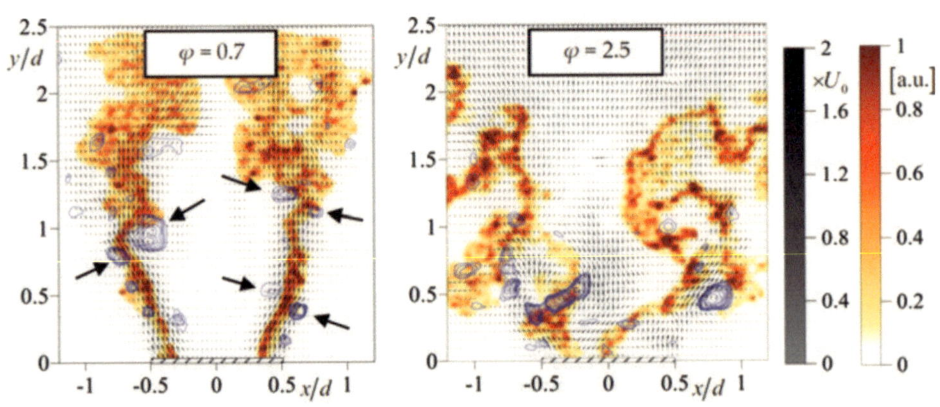

Figure 11. The instantaneous velocity and HCHO PLIF snapshots for turbulent high-swirl jets ($S = 1.0$) with combustion. Large-scale vortical structures are visualized by regions with positive Q-criterion ($Q^{2D} \geq 6U_0^2d^{-2}$ with the step of $3U_0^2d^{-2}$).

http://dx.doi.org/10.5772/intechopen.79896

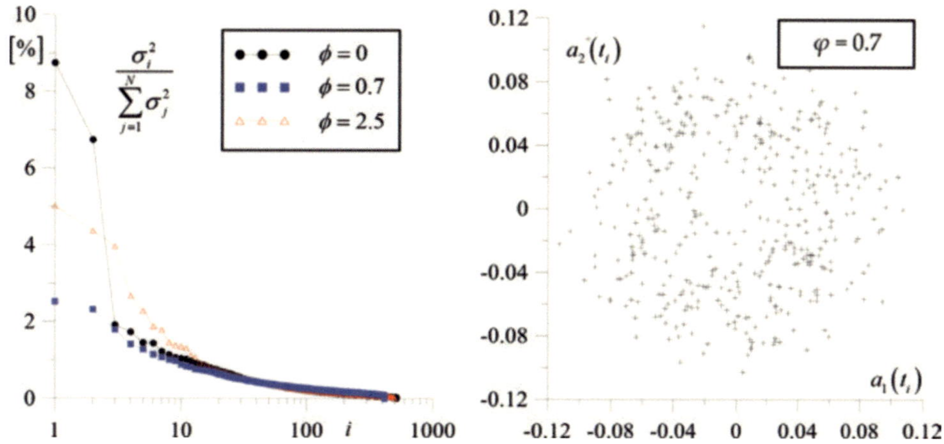

Figure 12. POD spectra for high-swirl jets (S = 1.0) with combustion and temporal coefficients for the first two POD modes for the fuel-lean flame.

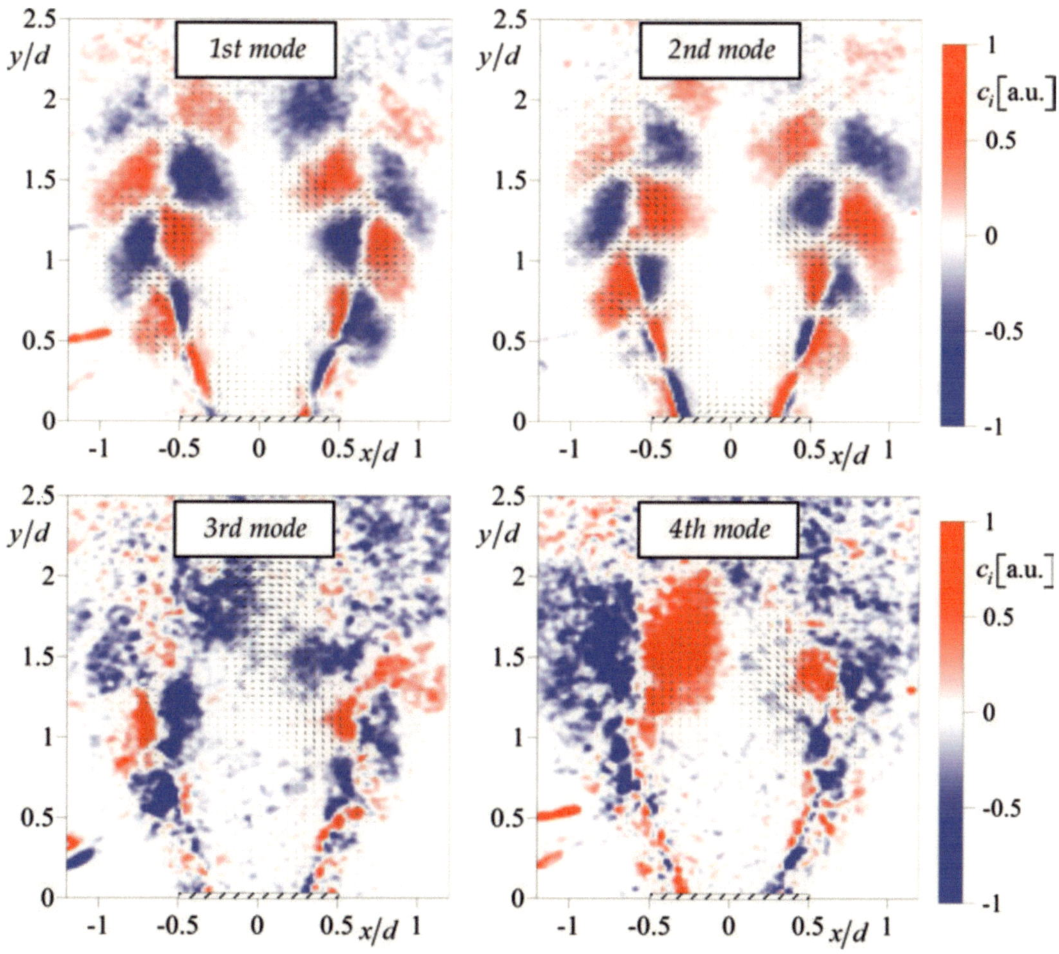

Figure 13. First four POD modes of the velocity fluctuations and conditionally sampled HCHO PLIF fluctuations for a high-swirl fuel-lean flame (S = 1.0, φ = 0.7).

As for the nonreacting flow, the first two POD modes for the fuel-lean flame correspond to coherent velocity fluctuations in the inner and outer mixing layers. The coherent fluctuations of HCHO data correspond to traveling waves along the flame front, which spread downstream. This correlation is related to the regular flame front deformations during propagation of the large-scale vortex structures. The third and fourth POD modes correspond to the velocity fluctuations in the upper part of the recirculation zone.

For the fuel-rich flame, the spatial structure of the first three POD modes is similar to that reported in [37] for the similar kind of flame, whereas amplitudes of these modes are different. The latter finding is explained by the different field of view in the present and previous experiments. Despite coherent velocity fluctuations in the mixing layers, correlated with the transverse flow movement at the nozzle exit, strong, almost axisymmetric variation of the longitudinal velocity takes place in the outer mixing later, induced by the buoyancy force and resulted in an unsteady entrainment of surrounding air.

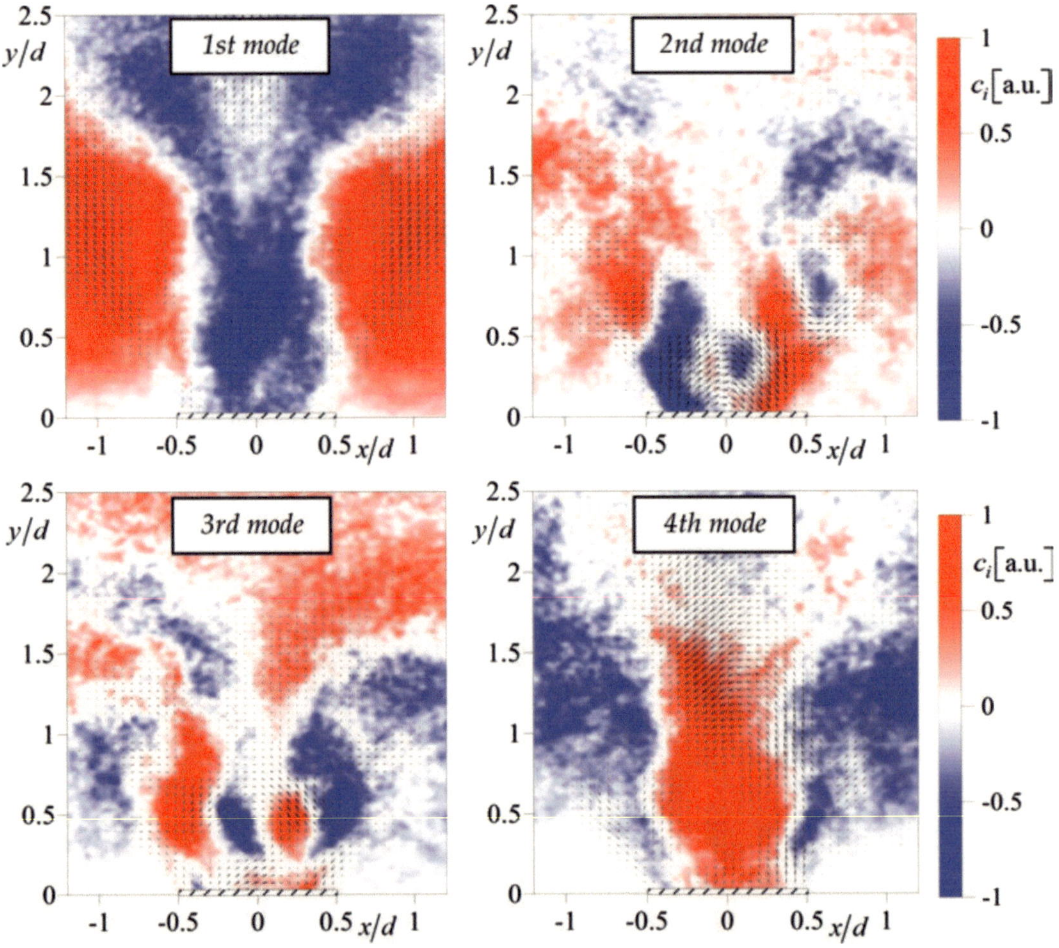

Figure 14. First four POD modes of the velocity fluctuations and conditionally sampled HCHO PLIF fluctuations for a high-swirl fuel-rich flame ($S = 1.0$, $\varphi = 2.5$).

4. Conclusions

Contribution of large-scale coherent structures on regular patterns during turbulent mixing in jet flows with different swirl rates is evaluated on the basis of PIV/PLIF measurements and POD of the velocity snapshots. The main findings are the following:

- For the nonswirling and weakly swirling jet (without permanently present central recirculation zone), toroidal vortices in the outer mixing layer are associated with regular patterns of the concentration fluctuations. For the low-swirl jet, the POD has also revealed variation of the axial velocity in the wake region and transverse movement of the vortex core. Such flow dynamics supports the unsteady vortex breakdown scenario of Oberleithner et al. [12] and is found to be correlated with alternation of the entrainment rate downstream.
- For the strongly swirling jets with vortex breakdown and central recirculation zone, the flow dynamics is related to rotation of a coherent structure, consisting of a pair of large-scale helical vortices. Such coherent structure remains in the considered cases of the fuel-lean and fuel-rich flames. The helical vortices surround the flame front of the lean mixture from both sides and provide regular deformations and stretching of the flame. The outer helical vortex produces ordered concentration structures during the mixing of the jet with the surrounding air and causes large-scale coherent motion of the fuel-rich flame.

Acknowledgements

The research is supported by the Russian Science Foundation (Grant № 16-19-10566). The authors are grateful to Prof. K. Hanjalic for his fruitful discussions and recommendations.

Conflict of interest

The authors declare no conflict of interest.

Author details

Vladimir Dulin[1,2]*, Aleksei Lobasov[1,2], Dmitriy Markovich[1,2] and Sergey Alekseenko[1,2]

*Address all correspondence to: vmd@itp.nsc.ru

1 Kutateladze Institute of Thermophysics, Siberian Branch of the Russian Academy of Sciences, Novosibirsk, Russia

2 Novosibirsk State University, Novosibirsk, Russia

References

[1] Gupta AK, Lilley DG, Syred N. Swirl Flows. Kent: Abacus Press; 1984. 488 p. DOI: 10.1016/0010-2180(86)90133-1

[2] Fröhlich J, García-Villalba M, Rodi W. Scalar mixing and large-scale coherent structures in a turbulent swirling jet. Flow, Turbulence and Combustion. 2008;**80**(1):47-59. DOI: 10.1007/s10494-007-9121-3

[3] Cozzi F, Coghe A, Sharma R. Analysis of local entrainment rate in the initial region of a isothermal free swirling jets by stereo PIV. Experimental Thermal and Fluid Science. 2018; **94**:281-294. DOI: 10.1016/j.expthermflusci.2018.01.013

[4] Zemtsop CP, Stöllinger MK, Heinz S, Stanescu D. Large-eddy simulation of swirling turbulent jet flows in absence of vortex breakdown. AIAA Journal. 2009;**47**(12):3011-3021. DOI: 10.2514/1.43813

[5] McIlwain S, Pollard A. Large eddy simulation of the effects of mild swirl on the near field of a round free jet. Physics of Fluids. 2002;**14**(2):653-661. DOI: 10.1063/1.1430734

[6] Loiseleux T, Chomaz JM. Breaking of rotational symmetry in a swirling jet experiment. Physics of Fluids. 2003;**15**(2):511-523. DOI: 10.1063/1.1533068

[7] Liang H, Maxworthy T. An experimental investigation of swirling jets. Journal of Fluid Mechanics. 2005;**525**:115-159. DOI: 10.1017/S0022112004002629

[8] Markovich DM, Dulin VM, Abdurakipov SS, Kozinkin LA, Tokarev MP, Hanjalić K. Helical modes in low-and high-swirl jets measured by tomographic PIV. Journal of Turbulence. 2016;**17**(7):678-698. DOI: 10.1080/14685248.2016.1173697

[9] Billant P, Chomaz JM, Huerre P. Experimental study of vortex breakdown in swirling jets. Journal of Fluid Mechanics. 1998;**376**:183-219. DOI: 10.1017/S0022112098002870

[10] Lucca-Negro O, O'doherty T. Vortex breakdown: A review. Progress in Energy and Combustion Science. 2001;**27**(4):431-481. DOI: 10.1016/S0360-1285(00)00022-8

[11] Ruith MR, Chen P, Meiburg E, Maxworthy T. Three-dimensional vortex breakdown in swirling jets and wakes: Direct numerical simulation. Journal of Fluid Mechanics. 2003; **486**:331-378. DOI: 10.1017/S0022112003004749

[12] Oberleithner K, Paschereit CO, Seele R, Wygnanski I. Formation of turbulent vortex breakdown: Intermittency, criticality, and global instability. AIAA Journal. 2012;**50**(7):1437-1452. DOI: 10.2514/1.J050642

[13] Syred N, Chigier NA, Beér JM. Flame stabilization in recirculation zones of jets with swirl. Proceedings of the Combustion Institute. 1971;**13**:617-624. DOI: 10.1016/S0082-0784(71)80063-2

[14] Mourtazin D, Cohen J. The effect of buoyancy on vortex breakdown in a swirling jet. Journal of Fluid Mechanics. 2007;**571**:177-189. DOI: 10.1017/S0022112006002862

[15] Hartung G, Hult J, Kaminski CF, Rogerson JW, Swaminathan N. Effect of heat release on turbulence and scalar-turbulence interaction in premixed combustion. Physics of Fluids. 2008;**20**(3):035110. DOI: 10.1063/1.2896285

[16] Tangirala V, Driscoll JF. Temperatures within non-premixed flames: Effects of rapid mixing due to swirl. Combustion Science and Technology. 1988;**60**(1–3):143-162. DOI: 10.1080/00102208808923981

[17] Bradley D, Gaskell PH, Gu XJ, Lawes M, Scott MJ. Premixed turbulent flame instability and NO formation in a lean-burn swirl burner. Combustion and Flame. 1998;**115**(4):515-538. DOI: 10.1016/S0010-2180(98)00024-8

[18] Schmittel P, Günther B, Lenze B, Leuckel W, Bockhorn H. Turbulent swirling flames: Experimental investigation of the flow field and formation of nitrogen oxide. Proceedings of the Combustion Institute. 2000;**28**(1):303-309. DOI: 10.1016/S0082-0784(00)80224-6

[19] Meier W, Weigand P, Duan XR, Giezendanner-Thoben R. Detailed characterization of the dynamics of thermoacoustic pulsations in a lean premixed swirl flame. Combustion and Flame. 2007;**150**(1–2):2-26. DOI: 10.1016/j.combustflame.2007.04.002

[20] Meier W, Dem C, Arndt CM. Mixing and reaction progress in a confined swirl flame undergoing thermo-acoustic oscillations studied with laser Raman scattering. Experimental Thermal and Fluid Science. 2016;**73**:71-78. DOI: 10.1016/j.expthermflusci.2015.09.011

[21] Konle M, Kiesewetter F, Sattelmayer T. Simultaneous high repetition rate PIV–LIF measurements of CIVB driven flashback. Experiments in Fluids. 2008;**44**(4):529-538. DOI: 10.1007/s00348-007-0411-2

[22] Konle M, Sattelmayer T. Interaction of heat release and vortex breakdown during flame flashback driven by combustion induced vortex breakdown. Experiments in Fluids. 2009; **47**:627-635. DOI: 10.1007/s00348-009-0679-5

[23] Tangermann E, Pfitzner M. Evaluation of combustion models for combustion-induced vortex breakdown. Journal of Turbulence. 2009;**10**(7):1-21. DOI: 10.1080/14685240802592423

[24] Syred N, Beer J. The damping of precessing vortex cores by combustion in swirl generators. Astronautica Acta. 1972;**17**(4):783-801

[25] Schneider C, Dreizler A, Janicka J. Fluid dynamical analysis of atmospheric reacting and isothermal swirling flows. Flow, Turbulence and Combustion. 2005;**74**(3):103-127. DOI: 10.1007/s10494-005-7369-z

[26] Duwig C, Fuchs L. Large eddy simulation of vortex breakdown/flame interaction. Physics of Fluids. 2007;**19**:075103. DOI: 10.1063/1.2749812

[27] Nogenmyr KJ, Fureby C, Bai XS, Petersson P, Collin R, Linne M. Large eddy simulation and laser diagnostic studies on a low swirl stratified premixed flame. Combustion and Flame. 2009;**155**(3):25-36. DOI: 10.1016/j.combustflame.2008.06.014

[28] Day M, Tachibana S, Bell J, Lijewski M, Beckner V, Cheng RK. A combined computational and experimental characterization of lean premixed turbulent low swirl laboratory flames: I. Methane flames. Combustion and Flame. 2012;**159**(1):275-290. DOI: 10.1016/j.combustflame.2011.06.016

[29] Day M, Tachibana S, Bell J, Lijewski M, Beckner V, Cheng RK. A combined computational and experimental characterization of lean premixed turbulent low swirl laboratory flames II. Hydrogen flames. Combustion and Flame. 2015;**162**(5):2148-2165. DOI: 10.1016/j.combustflame.2015.01.013

[30] Mansouri Z, Aouissi M, Boushaki T. Numerical computations of premixed propane flame in a swirl-stabilized burner: Effects of hydrogen enrichment, swirl number and equivalence ratio on flame characteristics. International Journal of Hydrogen Energy. 2016; **41**(22):9664-9678. DOI: 10.1016/j.ijhydene.2016.04.023

[31] Mansouri Z, Aouissi M, Boushaki T. Detached eddy simulation of high turbulent swirling reacting flow in a premixed model burner. Combustion Science and Technology. 2016;**188** (11–12):1777-1798. DOI: 10.1080/00102202.2016.1211888

[32] Lieuwen TC. Unsteady Combustor Physics. Cambridge: Cambridge University Press; 2012. 426 p

[33] Syred N. A review of oscillation mechanisms and the role of the precessing vortex core (PVC) in swirl combustion systems. Progress in Energy and Combustion Science. 2006; **32**(2):93-161. DOI: 10.1016/j.pecs.2005.10.002

[34] Cala CE, Fernandes E, Heitor MV, Shtork SI. Coherent structures in unsteady swirling jet flow. Experiments in Fluids. 2006;**40**(2):267-276. DOI: 10.1007/s00348-005-0066-9

[35] Legrand M, Nogueira J, Lecuona A, Nauri S, Rodríguez PA. Atmospheric low swirl burner flow characterization with stereo PIV. Experiments in Fluids. 2010;**48**(5):901-913. DOI: 10.1007/s00348-009-0775-6

[36] Oberleithner K, Sieber M, Nayeri CN, Paschereit CO, Petz C, Hege HC, Noack BR, Wygnanski I. Three-dimensional coherent structures in a swirling jet undergoing vortex breakdown: Stability analysis and empirical mode construction. Journal of Fluid Mechanics. 2011;**679**:383-414. DOI: 10.1017/jfm.2011.14

[37] Alekseenko SV, Dulin VM, Kozorezov YS, Markovich DM. Effect of high-amplitude forcing on turbulent combustion intensity and vortex core precession in a strongly swirling lifted propane/air flame. Combustion Science and Technology. 2012;**184**(10–11):1862-1890. DOI: 10.1080/00102202.2012.695239

[38] Stöhr M, Boxx I, Carter CD, Meier W. Experimental study of vortex-flame interaction in a gas turbine model combustor. Combustion and Flame. 2012;**159**(8):2636-2649. DOI: 10.1016/j.combustflame.2012.03.020

[39] Markovich DM, Abdurakipov SS, Chikishev LM, Dulin VM, Hanjalić K. Comparative analysis of low-and high-swirl confined flames and jets by proper orthogonal and dynamic mode decompositions. Physics of Fluids. 2014;**26**(6):065109. DOI: 10.1063/1.4884915

[40] Ceglia G, Discetti S, Ianiro A, Michaelis D, Astarita T, Cardone G. Three-dimensional organization of the flow structure in a non-reactive model aero engine lean burn injection system. Experimental Thermal and Fluid Science. 2014;**52**:164-173. DOI: 10.1016/j.expthermflusci.2013.09.007

[41] Boxx I, Stöhr M, Carter C, Meier W. Temporally resolved planar measurements of transient phenomena in a partially pre-mixed swirl flame in a gas turbine model combustor. Combustion and Flame. 2010;**157**(8):1510-1525. DOI: 10.1016/j.combustflame.2009.12.015

[42] Stöhr M, Sadanandan R, Meier W. Phase-resolved characterization of vortex–flame interaction in a turbulent swirl flame. Experiments in Fluids. 2011;**51**(4):1153-1167. DOI: 10.1007/s00348-011-1134-y

[43] Glassman I. Combustion. 3rd ed. San Diego: Academic Press; 1996. 631 p

[44] Brackmann C, Nygren J, Bai X, Li Z, Bladh H, Axelsson B, Denbratt I, Koopmans L, Bengtsson P-E, Alden M. Laser-induced fluorescence of formaldehyde in combustion using third harmonic Nd:YAG laser excitation. Spectrochimica Acta Part A. 2003;**59**: 3347-3356. DOI: 10.1016/S1386-1425(03)00163-X

[45] Syred N, Beer JM. Combustion in swirling flows: A review. Combustion and Flame. 1974; **23**(2):143-201. DOI: 10.1016/0010-2180(74)90057-1

[46] Anacleto PM, Fernandes EC, Heitor MV, Shtork SI. Swirl flow structure and flame characteristics in a model lean premixed combustor. Combustion Science and Technology. 2003; **175**(8):1369-1388. DOI: 10.1080/00102200302354

[47] Fernandes EC, Heitor MV, Shtork SI. An analysis of unsteady highly turbulent swirling flow in a model vortex combustor. Experiments in Fluids. 2006;**40**(2):177-187. DOI: 10.1007/s00348-005-0034-4

[48] Scarano F. Iterative image deformation methods in PIV. Measurement Science and Technology. 2001;**13**(1):R1-R19. DOI: 10.1088/0957-0233/13/1/201

[49] Coudert SJM, Schon J-P. Back-projection algorithm with misalignment corrections for 2D3C stereoscopic PIV. Measurement Science and Technology. 2001;**12**:1371-1381. DOI: 10.1088/0957-0233/12/9/301

[50] Wang GH, Clemens NT. Effects of imaging system blur on measurements of flow scalars and scalar gradients. Experiments in Fluids. 2004;**37**:194-205. DOI: 10.1007/s00348-004-0801-7

[51] Sirovich L. Turbulence and the dynamics of coherent structures. I. Coherent structures. Quarterly of Applied Mathematics. 1987;**45**(3):561-571

[52] Hunt JCR, Wray AA, Moin P. Eddies, stream, and convergence zones in turbulent flows. CTR Report S88. 1988

[53] Broze G, Hussain F. Transitions to chaos in a forced jet: Intermittency, tangent bifurcations and hysteresis. Journal of Fluid Mechanics. 1996;**311**:37-71. DOI: 10.1017/S002211209600 2509

Chapter 3

Hydrodynamic Vortex Structures in a Diffusion Jet Flame

Dubnishchev Yuri Nikolaevich,
Lemanov Vadim Vladimirovich,
Lukashov Vladimir Vladimirovich,
Arbuzov Vitali Anisiforovich and
Sharov Konstantin Aleksandrovich

Additional information is available at the end of the chapter

http://dx.doi.org/10.5772/intechopen.80610

Abstract

The chapter presents the results of an experimental study of hydrodynamics and diffusion combustion of jets flowing out of long tubes in the Reynolds number range 200–13,500 into air. The methods used in the experiments are visualization in the ultraviolet region, PIV, and hot-wire anemometry. The amplitude-phase structure of optical filters in systems of the Hilbert diagnostics of phase optical density fields in gaseous and condensed systems was used in this work. A possibility of visualization of disturbances in phase optical density fields with arbitrary amplitudes is demonstrated. Two geometries are studied: jet combustion in a stationary atmosphere and in a cross flow. Propane and hydrogen in a mixture with an inert diluent (CO_2) were used as fuels. In the isothermal jet stream, propane-butane mixture and Freon-22 are used. The main attention in the problem is paid to the mechanism of pipe and jet instability interaction, resulting in the vortex motion in several spatial regions. For critical Reynolds numbers in a pipe, the characteristic is the mechanism of two-stage instability caused by turbulent spot (puff) formation inside the pipe and vortex structure generation in the jet-mixing layer. These vortex structures (puff) exert a strong influence both on the isothermal jet and on the flame.

Keywords: jet flame, combustion control, laminar-turbulent transition, vortex structures, turbulence, experiment, visualization

1. Introduction

At present, there are many detailed studies of the jets flowing out of contoured nozzles for turbulent flow regimes (Reynolds number over 5000), where the Kelvin-Helmholtz instability is the key mechanism for the formation of vortices in the initial region [1–3]. The diffusion torch has been also fairly well studied [4]. One of the new and promising methods for controlling combustion is the jet diffusion flame in the regimes of instability evolving both in the jet stream itself and inside the jet source [5–7]. A feature of this problem is the use of jets with low Reynolds numbers, implying low speeds and low mass flow fuel rates. It agrees with the urgent tasks of energy saving and energy efficiency. Such regimes allow obtaining jets with a long laminar zone length. Two-stage instability (inside the jet source and in the jet-mixing layer) may serve as a mechanism controlling the jet flame organization.

The principal difference of this problem statement compared to other works is the mechanism of interaction between instabilities in tube and jet. This leads to the organization of a vortex motion in several spatial regions. At critical Reynolds numbers in a pipe through which fuel is supplied, local organized vortex structures of two types (puff and slugs) are formed [8]. In the jet part of the flame, it is possible to form several vortex zones. A low-frequency flicker instability occurs at the outer boundary of the flame, and the shear instability of the mixing layer can be observed in the core of the torch (in the near-axis zone) [6, 9, 10]. Vortex structures (puff and slugs) formed in transient flow regimes in the tube can have a significant effect on combustion. As shown by our experiments [7], the range of regulation of the flame structure at low Reynolds numbers (Re = 1800–3000) is quite wide. With the help of puff or slugs, it is possible to initiate a transition from a laminar flame to a turbulent one, or vice versa, to realize a laminarization of the reacting flow. As is known, the flowing from the tube (nozzle) can form two types of flame—attached and detached (lifted flame). Using the mechanism of forming vortex perturbations inside the jet source, one can switch the flame from one type to another, and vice versa. And in the case of the action of a vortex disturbance of large amplitude, even a flame blowoff is possible.

The chapter presents the results of an experimental study of jets and diffusion burning of jets by the example of an outflow from a long tube at low Reynolds numbers. Two geometries are studied: jet combustion in a stationary atmosphere and in a transverse airflow. To achieve the goal, the following methods are used: Hilbert visualization, PIV, and hot-wire anemometry. During combustion, propane and hydrogen are used as a fuel in the mixture with an inert gas. Isothermal jets of various gases (propane-butane, CO_2, Freon-22) are also considered. The experiments are carried out when a subsonic gas jet flows into the air space from a tube of 2 and 3.2 mm diameter in the Reynolds number range of 200–15,000. At that, the speed range is 0.3–60 m/s.

2. Optical methods of diagnostics

Optical diagnostics of flows have been successfully applied in experimental hydrodynamics and gas dynamics for a long time. It offers a large variety of modern methods and techniques: shadowgraphy and Schlieren methods [11–13], laser Doppler anemometry [13, 14], Doppler technologies of velocity field measurements [15, 16], and particle image velocimetry [17]. The Hilbert diagnostics of flows [18–21] based on visualization of space-and-time phase disturbances induced in the light field by the flow passing through the examined medium has a high potential for development and application. Such disturbances are visualized by means of the Hilbert filtration of the optical field, transforming the latter into a Hilbert-conjugate signal.

If the Foucault-Hilbert transform is performed, the result is an analytical signal being superposition of the initial and Hilbert-conjugate fields. The amplitude of these signals contains information about the structure of the phase disturbance.

Though the theoretical and experimental details of the Hilbert optics have been discussed in many publications, its potential is far from being exhausted. This refers, in particular, to the analysis of the amplitude-phase characteristics of the filters and the dynamic range of phase disturbances during the Hilbert diagnostics in the spectral range of the probing field.

The optical measuring complex is based on the serial IAB-463 M shadow device [www.skb-photon] using the optical Hilbert filtering modules, shear interferometry, and light source, specially adapted for the purpose of the experiment. For interferometric studies, a modified shear interferometer is used [18]. **Figure 1** shows a simplified scheme of the experimental setup consisting of an experimental stand and an optical measuring complex. The circuit contains a lighting module consisting of a light source 1, a collimator lens 2, and a slit diaphragm 3. The slit diaphragm is located in the front focal plane of the objective 4, forming a light probe field in the medium under study. The Fourier spectrum of the phase perturbations induced in the medium under investigation is formed by an objective 5 in the frequency plane where a quadrant Hilbert filter 6 is placed. The objective 7 performs the inverse Fourier transformation of the filtered light field, visualizing its phase perturbations, which are recorded by a digital video camera 8 connected to the computer 9.

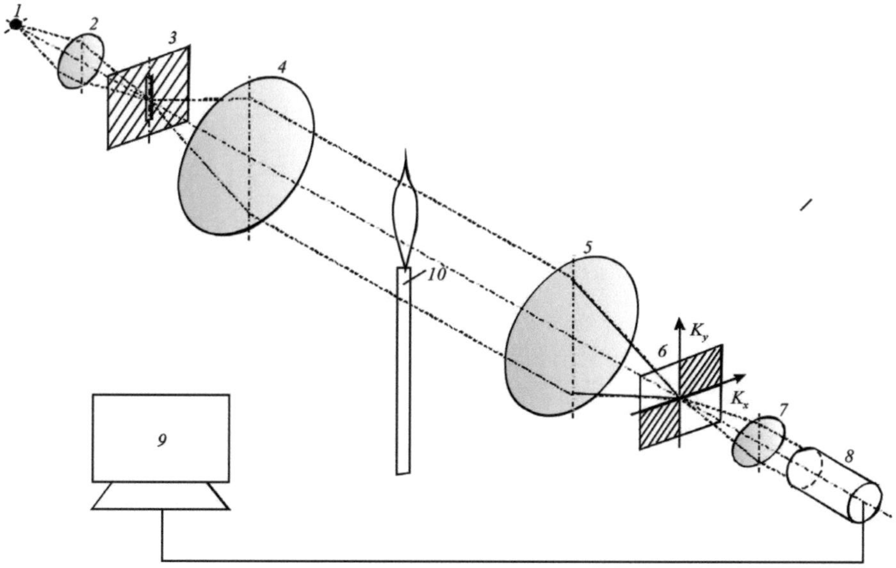

Figure 1. Experimental scheme.

The coherent transfer function of the spatial-frequency filter (HF) performing a one-dimensional Hilbert transform is described by the expression

$$H(K_x, K_y) = \left[e^{-i\varphi}\sigma(K_x) + e^{i\varphi}\sigma(-K_x)\right]\sigma(K_y) + \left[e^{i\varphi}\sigma(K_x) + e^{-i\varphi}\sigma(-K_x)\right]\sigma(-K_y), \quad (1)$$

where K_x and K_y are the spatial frequencies, $\sigma(\pm K_x)$ and $\sigma(\pm K_y)$ are the Heaviside functions, and φ is the phase shift defined by the corresponding quadrant of the spatial-frequency filter. Given that $\sigma(\pm K_x) = \frac{1}{2}[1 \pm \text{sgn}K_x]$, where $\text{sgn}K_x$ is the sign function, expression Eq. (1) for the coherent transfer function of the filter takes the form

$$\begin{aligned} H(K_x, K_y) &= (\cos\varphi - i\sin\varphi\,\text{sgn}\,K_x)\sigma(K_y) + (\cos\varphi + i\sin\varphi\,\text{sgn}\,K_x)\sigma(-K_y) \\ &= \cos\varphi - i\left[\sigma(K_y) - \sigma(-K_y)\right]\sin\varphi\,\text{sgn}\,K_x. \end{aligned} \quad (2)$$

The dynamic representation of the Heaviside functions is used as follows:

$$\sigma(K_y) - \sigma(-K_y) = \int_{-\infty}^{\infty} \sigma(\xi)\delta(K_y - \xi)d\xi - \int_{-\infty}^{\infty} \sigma(\xi)\delta(K_y + \xi)d\xi. \quad (3)$$

Variable ξ is the projection of an arbitrary point of the slit source onto the spatial-frequency axis K_y. This projection has the physical meaning of the zero frequency reference K_x on the spatial-frequency axis K_y. Since $\sigma(\xi) = \frac{1}{2}[1 + \text{sgn}\,\xi]$, expression Eq. (3) takes the form

$$\begin{aligned} \sigma(K_y) - \sigma(-K_y) &= \frac{1}{2}\left\{\int_{-\infty}^{\infty}[1 + \text{sgn}\,\xi]\delta(K_y - \xi)d\xi - \int_{-\infty}^{\infty}[1 - \text{sgn}\,\xi]\delta(K_y + \xi)d\xi\right\} \\ &= \frac{1}{2}\left\{\int_{-\infty}^{\infty}\delta(K_y - \xi)d\xi + \int_{-\infty}^{\infty}\text{sgn}\,\xi\delta(K_y - \xi)d\xi - \int_{-\infty}^{\infty}\delta(K_y + \xi)d\xi + \int_{-\infty}^{\infty}\text{sgn}\,\xi\delta(K_y + \xi)d\xi\right\} \\ &= \frac{1}{2}\{1 + \text{sgn}\,K_y + 1 + \text{sgn}(-K_y)\} = 1. \end{aligned}$$

Substitution of $\sigma(K_y) - \sigma(-K_y) = 1$ into Eq. (2)

$$H(K_x, K_y) = \cos\varphi - i\sin\varphi\,\text{sgn}\,K_x. \quad (4)$$

The filter with the coherent transfer function (Eq. (4)) performs a one-dimensional Foucault-Hilbert transform.

In the Fourier plane $H(K_x, K_y)$, the spatial-frequency axis K_x is orthogonal to the image of the slit source. The Fourier spectrum of the light field immediately after the filter has the form

$$s(K_x, K_y)H(K_x, K_y) = s(K_x, K_y)[(\cos\varphi - i\sin\varphi\,\text{sgn}\,K_x)] = s(K_x, K_y)\cos\varphi + \hat{s}_x(K_x, K_y)\sin\varphi. \quad (5)$$

where $s(K_x, K_y)$ is the spatial-frequency Fourier spectrum of the light field perturbed by the test medium; $\hat{s}_x(K_x, K_y) = -i\text{sgn}(K_x)s(K_x, K_y)$ is the Fourier spectrum of the light field subjected to the one-dimensional Hilbert transform on the K_x axis.

The phase shift φ is a function of the wavelength λ of the probing light field: $\varphi = \varphi(\lambda)$. At a wavelength $\lambda = \lambda_0$ that satisfies the condition $\varphi(\lambda_0) = \pi/2$, the coherent transfer function (Eq. (4)) takes the form

$$H(K_x, K_y) = -i\,\mathrm{sgn}\,K_x. \tag{6}$$

In this case, the Fourier filter HF performs a one-dimensional Hilbert transform:

$$\hat{s}_x(K_x, K_y) = -i\,\mathrm{sgn}\,K_x s(K_x, K_y). \tag{7}$$

If the investigated medium induces only phase perturbations of the probing field, the Fourier spectrum of the perturbed field is given by

$$s(K_x, K_y) = e^{i\psi(K_x, K_y)}.$$

Accordingly, for the Fourier spectrum of the filtered field, the following is obtained

$$\hat{s}(K_x, K_y) = e^{i\psi(K_x, K_y)} H(K_x, K_y). \tag{8}$$

As follows from Eq. (8), the Fourier spectrum of phase perturbations has a structure consisting of isophase lines satisfying the equation.

$$\psi(K_x, K_y) = \pi m,$$

where $m = 1,\ 2,\ 3\ldots$ and then $\tilde{\psi}(K_x, K_y)$ takes the value of a perturbation of the isophase line that broadens it:

$$\psi = \pi m + \tilde{\psi}(K_x, K_y), \tilde{\psi} << 1.$$

In this case, the Fourier spectrum of the light field perturbed by the test medium obtains

$$s(K_x, K_y) = \sum_m e^{i\left[\pi m + \tilde{\psi}_m(K_x, K_y)\right]} \approx \sum_m (-1)^m e^{i\tilde{\psi}_m(K_x, K_y)},$$

or taking into account that $\tilde{\psi}_m(K_x, K_y) << 1$,

$$s(K_x, K_y) = \sum_m (-1)^m \left[1 + i\tilde{\psi}_m(K_x, K_y)\right]. \tag{9}$$

In view of Eq. (9), the filtered Fourier spectrum of the phase perturbations is expressed as

$$s(K_x, K_y)H(K_x, K_y) = \left\{\sum_m (-1)^m \left[1 + i\tilde{\psi}_{m_x}(K_x, K_y)\right]\right\}[-i\,\mathrm{sgn}\,K_x] = \sum_m (-1)^m \hat{\psi}_{m_x}(K_x, K_y), \tag{10}$$

where $\hat{\psi}_{m_x}(K_x, K_y)$ is the Fourier spectrum of the phase perturbations visualized by means of the Hilbert transform on the x axis. As follows from Eq. (10), phase structures are visualized by

Hilbert filtering. Here, it is taken into account that the Hilbert image of constant quantities is zero according to the properties of the Hilbert transform.

The camcorder lens $L4$ performs an inverse Fourier transform of the filtered Fourier spectrum of phase perturbations:

$$\sum_m \hat{\psi}_{m_x}(K_x, K_y) \leftrightarrow \sum_m \hat{\psi}_{m_x}(x, y). \tag{11}$$

The phase structures Eq. (11) visualized by the one-dimensional Hilbert transform operation are recorded by the camcorder CCD array:

$$\left|\sum_m \hat{\psi}_{m_x}(x, y)\right|^2 = \sum_m \left|\hat{\psi}_{m_x}(x, y)\right|^2. \tag{12}$$

Suppose that white light filtering is performed, the test medium is dispersionless, and the Fourier spectra are achromatic. The quadrature-phase filter performing a Hilbert transform at the wavelength λ_0 is used for filtering in the wavelength range (λ_1, λ_2):

$$H(K_x, K_y) = \cos\varphi(\lambda) - i \sin\varphi(\lambda) \operatorname{sgn} K_x. \tag{13}$$

Integrating the filtered Fourier spectrum in the wavelength range (λ_1, λ_2)

$$\int_{\lambda_2}^{\lambda_2} s(K_x, K_y) H(K_x, K_y) d\lambda \approx s(K_x, K_y) \int_{\lambda_1}^{\lambda_2} H(K_x, K_y) d\lambda. \tag{14}$$

When using a white light source, the weight coefficients $\cos\varphi(\lambda)$ and $\sin\varphi(\lambda)$ (13) are transformed into

$$\int_{\lambda_1}^{\lambda_2} \cos\left(\frac{\pi}{2}\frac{\lambda}{\lambda_0}\right) d\lambda = \frac{4\lambda_0}{\pi} \cos\left[\frac{\pi}{4\lambda_0}(\lambda_2 + \lambda_1)\right] \sin\left[\frac{\pi}{4\lambda_0}(\lambda_2 - \lambda_1)\right]. \tag{15}$$

$$\int_{\lambda_1}^{\lambda_2} \sin\left(\frac{\pi}{2}\frac{\lambda}{\lambda_0}\right) d\lambda = \frac{4\lambda_0}{\pi} \sin\left[\frac{\pi}{4\lambda_0}(\lambda_1 + \lambda_2)\right] \sin\left[\frac{\pi}{4\lambda_0}(\lambda_1 - \lambda_2)\right]. \tag{16}$$

Hence we obtain the partial ratio of the Hilbert image to the original signal:

$$\frac{\int_{\lambda_1}^{\lambda_2} \sin(\lambda) d\lambda}{\int_{\lambda_1}^{\lambda_2} \cos(\lambda) d\lambda} = \operatorname{tg}\left[\frac{\pi}{4\lambda_0}(\lambda_1 + \lambda_2)\right] = \operatorname{tg}\left(\frac{\pi}{2} + \frac{\pi\Delta\lambda}{4\lambda_0}\right) = \operatorname{ctg}\left(\frac{\pi\Delta\lambda}{4\lambda_0}\right). \tag{17}$$

The weight coefficients $\cos\varphi(\lambda)$ and $\sin\varphi(\lambda)$ of the signal and a Hilbert-conjugate signal in the spectral band are equal: $\Delta\lambda = \lambda_1 - \lambda_2$. For $\Delta\lambda \to 0$, a pure Hilbert transform occurs. The

Hilbert transform has quasi-differentiation properties. Therefore, the extreme of the phase distribution and the region of the maximum phase gradient region are transformed into Hilbert ranges.

3. Experimental equipment

The first one studied in the work was the jet without combustion ("cold jet") and then the jet under combustion conditions ("hot jet"). In "cold" experiments, propane-butane mixture (50% propane and 50% butane) and Freon-22 (CHF_2Cl) were used as a working gas. In the experiments with flame, propane and hydrogen were used both in pure forms (Y = 99.9% by volume) and in the mixture with inert diluent (CO_2). The flow in the experiments had atmospheric pressure and the initial room temperature. The equipment included gas vessels, gas reducers, and two flowmeters (**Figure 2**). The gas flow rate and the composition of the fuel mixture were set using digital flowmeters El-Flow Bronkhorst. The viscosity of the gas mixture necessary for calculations was estimated using the Sutherland model in the Wilke approximation [22].

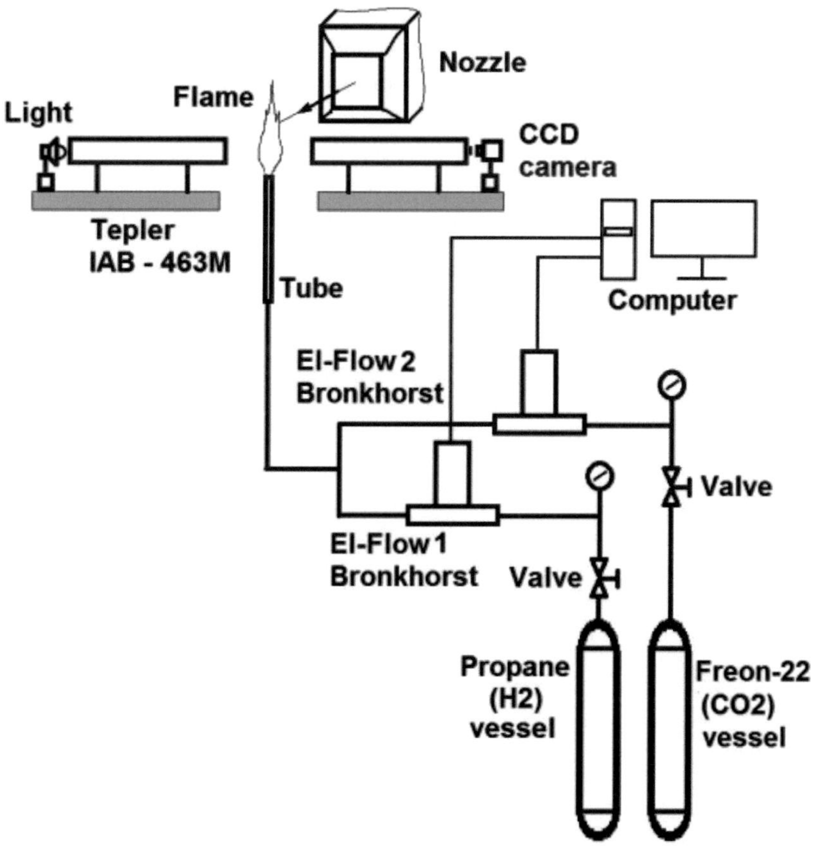

Figure 2. Scheme of setup.

Figure 2 shows a simplified scheme of the experimental setup with a complex for optical diagnostics. The registration of the shift interferograms and the video recording of the torch were made by a digital camera Canon 650D with a frequency of 50 fps. In addition, the high-speed visualization of light-scattering particles placed in the stream illuminated with a laser sheet was carried out with a Photron SA5 camera (4 kHz, the volume of one sample was 11,000 frames). The dynamic characteristics were measured by a constant temperature hot-wire anemometer DISA 55 M. To study the jet in a cross flow, a 2D PIV system was used; it is consisting of a digital camera with a 4 MPix matrix, a pulse laser with pulse energy of 90 mJ, a synchronizer, and a computer. To photograph the chemiluminescence of the OH radical (ultra-violet range), the DiCAM-PRO camera was used. Round tubes of quartz glass were used as a jet source (with an internal diameter d = 3.2 mm with a length L = 550 mm and d = 2 mm with a length L = 1000 mm). The transverse airflow was created with the help of an aerodynamic setup that allowed setting the speed at 1.5–15 m/s in the working part with a cross section of 100 × 100 mm.

4. Free jet investigation

At the first stage, experiments were performed in a "cold jet." It is known that the jet flow dynamics is significantly influenced by the initial conditions [1–3]. **Figure 3a** shows the dependence of the axial mean velocity (U_0) and the root-mean-square value of velocity pulsations (u) on the axis in the tube outlet cross-section as a function of the Reynolds number of the jet ($Re = d \cdot U_m/\nu$, where U_m is the bulk velocity). **Figure 3b** presents the dependence of the turbulence degree ($Tu = u/U_0 \times 100\%$) on Re. As can be seen from **Figure 3**, for the dependence of the average velocity on the Reynolds number, there is a local decrease in U_0 in the region of Re = 1800–2000. The experiments have shown that in this range of Re numbers, the velocity profile is changed from the laminar Poiseuille distribution to the fully developed turbulent profile.

For the root mean square value of the velocity pulsations u and the turbulence degree Tu, a local extremum is observed in the transition region at Re = 1920, and the maximum value of the

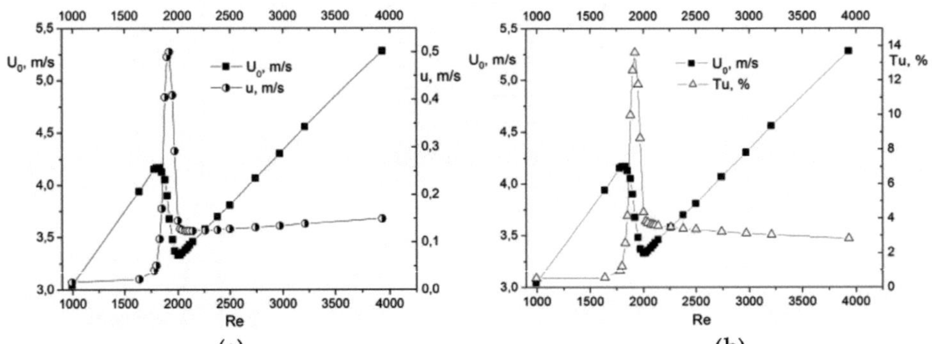

Figure 3. Parameters in the initial section on the axis of the "cold" jet (propane-butane) d = 3.2 mm and l = 500 mm: (a) the mean velocity and rms of the longitudinal velocity pulsations and (b) the mean velocity and degree of turbulence.

turbulence degree Tu = 13.5% is reached. The study of the oscillograms of the hot-wire anemometer signal at the tube outlet has shown the intermittent character of the instantaneous velocity behavior with respect to time and the presence of turbulent spots (puff) in the region of Reynolds transition numbers δ·Re/Re = 7–8%. At Re > 2100, the character of the oscillograms becomes completely turbulent. Thus, the obtained data indicate the presence of a laminar-turbulent transition at the outlet from the tube in the region of Re = 1800–2000 for the propane-butane mixture [8].

Further measurements were made by a hot-wire anemometer on the axis of the gas jet downstream. **Figure 4a** shows the dependence of the velocity on the jet axis U_x, normalized to the initial velocity U_0, on the dimensionless distance x/d for the Reynolds number variation (x is the longitudinal coordinate with its beginning at the start of the jet). The general character of the velocity variation along the axis is a monotonic decrease in the relative velocity U_x/U_0. As can be seen from **Figure 4**, the major changes of the velocity distribution along the axis occur in the region of Reynolds numbers characteristic for the laminar-turbulent transition Re = 1800–1915. At Re = 1820, the longest laminar flow zone is observed.

The change in the rms value of velocity pulsations and the degree of turbulence as a function of the Reynolds number on the jet axis are shown in **Figure 4b**. Here, several characteristic modes of the jet propagation can be distinguished. The instability waves in the mixing layer (basically, an asymmetric mode) were observed at Re > 200. For jets with small Reynolds numbers (Re = 500), an extended laminar flow zone with a low level of velocity pulsations is a characteristic. After that a laminar-turbulent transition region with an increase in pulsations (up to x/d = 20) takes place.

The next region is a zone of turbulent flow where pulsations decrease. With an increase in the Reynolds number, the length of the laminar part decreases, and the maximum of pulsations in the transition region increases. At Re = 1820, the length of the laminar zone is x/d = 10, and the pulsations level reaches Tu = 21%. At Re = 1915, there is a sharp increase in the level of velocity pulsations in the initial section of the jet. In the oscillogram of the hot-wire anemometer signal, an intermittent character of the behavior of instantaneous velocity with the presence of turbulent

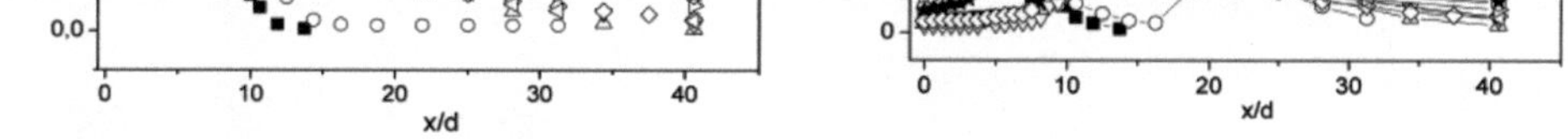

(a) (b)

Figure 4. Dynamic parameters along the jet axis with varying Re number, d = 3.2 mm L = 500 mm: (a) dimensionless mean velocity and (b) the degree of turbulence.

spots is observed. The above observations indicate that the laminar-turbulent transition occurs already inside the source of the jet (tube). As a result, the fluctuations in the near field of the jet increase sharply (at x/d = 7 to Tu = 30%). With a further increase in the Reynolds number (Re > 1915), the initial level of pulsations decreases to Tu_0 = 4%, and the maximum at x/d = 7 decreases to the value Tu = 12%. Further downstream (x/d > 7), there is a significant decrease in pulsations, which indicates a strong dissipation of turbulent energy.

Previously, experimental studies of the hydrodynamics of subsonic gas jets in the Reynolds number range 200–4000 have been carried out. One of the important issues is the length of the laminar part of the jet. Our research shows that it can be 100–250 d [23–25]. This task is promising from the point of view of the formation of a long-length flame and requires further study. In the "laminar" regime, when there is no puff, the instability of the jet flow prevails [2, 3], which allows controlling the vortex structures, for example, by acoustic action. The use of a denser gas as a working fluid makes it possible, due to the density gradient, to weaken the effect of this instability [26] and increase the length of the laminar part of the jet. Thus, it becomes possible to observe the development of a vortex structure in the jet stream.

The characteristic times for the existence of unstable flow regimes are characterized by the data presented in **Figure 5**. The experimental conditions correspond to **Figure 4b**; the sample was 11,000 frames; and the shooting frequency was 4 kHz. The observation time of the hydrodynamic disturbance in the field of the frame in the near field of the jet (x/d < 6) was taken as an event. The histogram of the vortex structure lifetime is distributed over time with a lag of 0.01 s. The ordinate represents the number of events in a given interval.

It can be seen that short events (the example is shown in **Figure 5a**) occur rarely (the first from the left quantile in **Figure 6**). Long-term events are also fixed quite seldom (the first from the

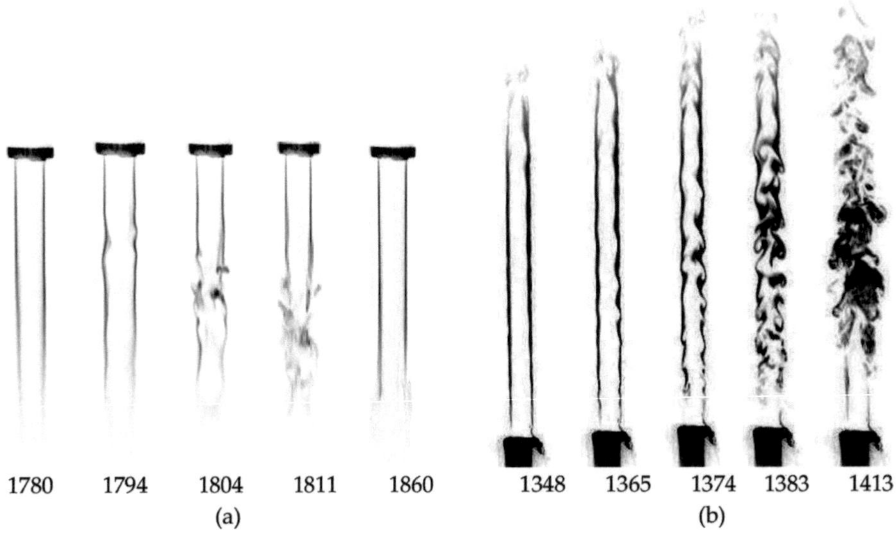

Figure 5. Expansion of the Freon-22 jet top-down (a) into the stationary air at Re = 1740 (see Video 1) and bottom-up (b) into the stationary air at Re = 2230 (Video 2).

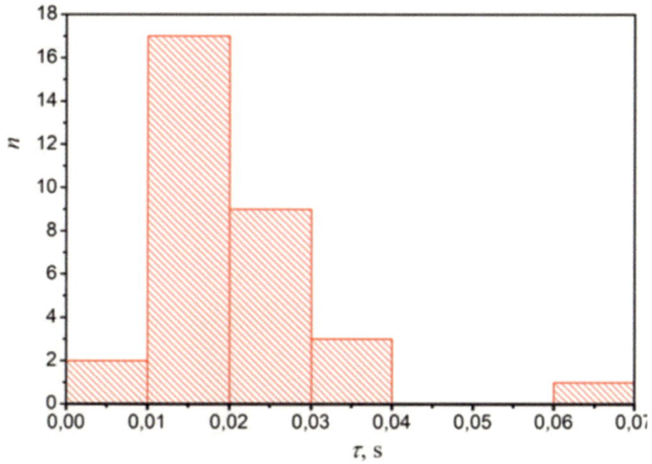

Figure 6. Histogram of the turbulent vortex structure distribution.

right quantile in **Figure 5**). The average time of the perturbation was 0.018 s (e.g., see **Figure 5b**). If the bulk velocity in the tube U_m is taken as a characteristic, then the estimate of the spatial scale of the jet instability in the near field gives the value L/d = 13.7. This value is close to the estimates of the extent of the turbulent spot inside the tube obtained by direct numerical simulation [26]. The question of the spatial vortex structure of such perturbations inside the tube is considered, for example, in [27, 28]. They represent a system of longitudinal vortices. Their arrangement in the cross section of the pipe has symmetry in the azimuth direction (a total of eight vortices). The presented investigations have shown (see **Figure 5**) that a turbulent spot in the initial part of the jet has a more complex vortex structure and requires additional studies.

A two-cascade instability mechanism is established in the work. The first cascade is a mode with intermittency inside the tube (alternation of the turbulent spots and laminar zones at the exit from the tube). The second cascade is the canonical instability of the jet-mixing layer. The study of the interaction of puff with large-scale structures of the mixing layer is the further task. This mechanism gives the prospect for controlling the flame by forming turbulent spots inside the pipe and vortex structures in the jet.

5. Investigating the diffusion jet flame

Quite many studies focus on the diffusion jet flame [4]. In this part of the paper, the main attention is paid to the interaction of vortex structures (puff), formed as a result of a laminar-turbulent transition in a long tube, with the jet flame. In this case, variants of the attached and lifted flame, as well as its extinction, are shown. The Reynolds number of the tube flow transition is conservative to the composition of the fuel mixture. However, the scenario diversity (very sensitive to the fuel mixture composition) of the flame-vortex interaction fundamentally distinguishes combustion from isothermal flow.

The main method of investigation in this part is Hilbert visualization. Diffusion combustion of a jet of a fuel mixture consisting of propane and an inert diluent (CO_2), flowing vertically upward into a stationary air atmosphere, illustrates the Hilbert visualization frames shown in **Figures 7** and **8** (d = 3.2 mm). The digits below each frame indicate the sequence number of the frame, and the shooting frequency is 50 fps. Just as in the case of the use of Freon-22, the experiments are presented in flow regimes, in which a laminar-turbulent transition with a characteristic intermittency process began inside the tube.

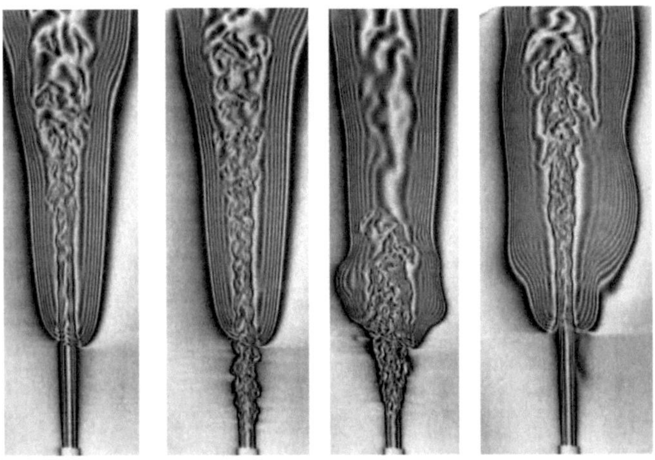

0530 0531 0532 0534

Figure 7. Laminar-lifted flame (Re = 2966, C_3H_8/CO_2, Y = 46).

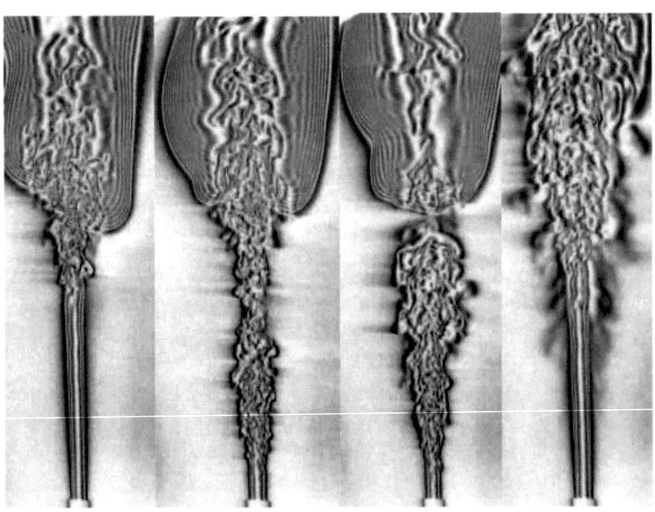

2190 2191 2192 2195

Figure 8. Flame extinction in the transient flow regime of the jet (Re = 2966, C_3H_8/CO_2, Y = 46) (see Video 3).

http://dx.doi.org/10.5772/intechopen.80610

In the attached mode of propane flame, the nature of the generation and dynamics of the propagation of the vortex structures along the flow do not undergo qualitative changes in comparison with the isothermal flow, although the influence of "puff-structures" on the low-frequency oscillations of the flame front is observed. When a puff appears in the observation zone, the wavelength of flicker oscillation increases by approximately 1.5–2 times. In the detached flame, the mutual influence of the vortex structures and the flame is much more pronounced. This may be caused by the known fact that in the vicinity of the flame front edge a local area of increased pressure is formed. **Figure 7** shows the effects of puff on a laminar-lifted flame. It can be seen that as the vortex structure passes along the jet, the leading edge of the flame undergoes significant changes in the spatial position and shape. In this case, low-frequency oscillations of the flame front leading edge are preserved, and combustion extinction is not observed. A completely different scenario of the flame-vortex interaction extinction is shown in **Figure 8**. The sequence of the Hilbert visualization frames in **Figure 8** illustrates the interaction of the vortex (frame 2191) with the lifted jet flame, which results in the extinction of the diffusion flame (frame 2195). The time interval between two adjacent frames (numbers) is 1/50 s.

Apparently, the impact of a large-scale perturbation led to the formation of a zone with a low-fuel content (frame 2192) in the near-axis area. These conditions are insufficient for combustion stabilization and led to a flame-off. So, vortex structure spontaneously arising in the tube affects the flame jet structure leading to its extinction.

It is known that the properties of hydrogen are very different from the thermal and chemical properties of hydrocarbons—H_2 has a low density and high values of the diffusion coefficient and, as a result, a high-flame propagation velocity. This inevitably affects the differences in the flame reaction to the appearance of vortex structures in the transient flow regime of the fuel in the tube. When hydrogen diluted with an inert gas is burned, the low-frequency flame flashes (flickering) can be suppressed almost completely (**Figure 9**).

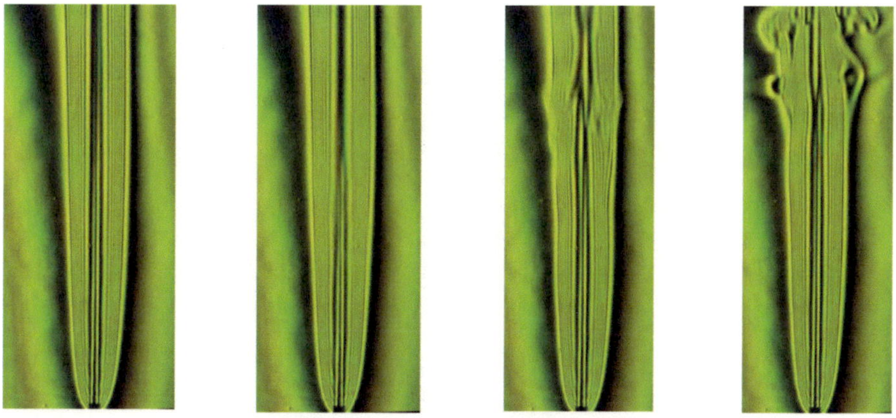

0143 0144 0145 0148

Figure 9. The attached H_2/CO_2 flame in the transient flow regime of the fuel jet (Re = 2870, Y = 90%) (see Video 4).

The flame can remain attached. The passage of vortex structures can be accompanied by the formation of "holes" in the flame—the dark areas shown in **Figure 9** (frame 0148), where combustion apparently, does not occur. Earlier, a similar phenomenon was discovered for the combustion of a methane jet at substantially higher Reynolds numbers [29].

6. The jet in a cross flow: a cold stream and a torch

Reacting jets are widely used in various types of burner devices. Jet flames have a simple, easily reproducible configuration and serve as an object of studies of the combustion dynamic fundamental principles, flame structure, and its stability both at combustion in still air and at interaction with air moving in different directions relative to the jet. One of the most important aspects of the problem is the flame extinction conditions, which determine the boundaries of the stable combustion and the range of admissible parameters for the stable operation of the burner device. Recent advances in the development of experimental diagnostics and numerical methods open new prospects in this field. For example, detailed experimental data on the flame structure for the combustion of a H_2/CO_2 mixture in a transverse flow, obtained using high-speed OH imaging and PLIF measurements, are presented in [30, 31], respectively.

One of the fundamental problems in the analysis of such processes is detecting and recording the flow parameters which ensure a stable combustion. Conditions for the jet flame extinction, i.e., the correlation of the cross flow velocity at which extinction occurs, fuel jet velocity, and other factors have been well established for hydrocarbon fuels. Recently, the conditions for stable burning of a jet flame of hydrogen fuel diluted with hydrocarbons (CO_2, CH_4, C_3H_8) in still air were investigated in [32]. The paper showed how additives of various gases affect the transition from the attached flame to the lifted one. In particular, it was noted that an increase in the proportion of hydrogen leads to a monotonic increase in the rate of the jet flame-off. In a number of studies, for example, in [33] it was noted that for detached and lifted flames, the extinction conditions differ significantly. In [34–36], semi-empirical methods were proposed to generalize the experimentally obtained data on the conditions of flame extinction in a cross

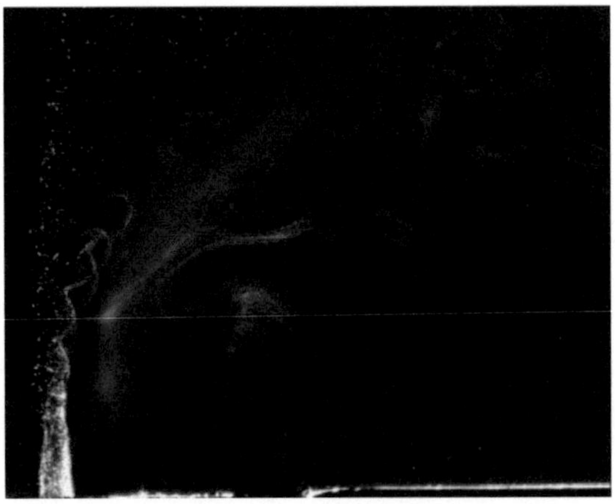

Figure 10. PIV visualization of the H_2/N_2 flame in air cross flow: d = 2 mm, H_2/N_2—Y = 36.7%, U_0 = 20 m/s, U_C = 8 m/s.

airflow. A review of physical mechanisms for jet flame stabilization is presented in [37]. Nevertheless, the parameters determining the hydrogen flame extinction for fuel mixtures with an inert diluent remain insufficiently investigated. To understand the mechanisms of turbulent flame stabilization, it is necessary to have detailed information on the structure and dynamics of heat-release regions in the flame and their interaction with the aerodynamics of the reacting flow. The topology of the heat-generating regions, even in the simplest configurations of the flow, may turn out to be nontrivial. Thus, for example, the PIV visualization of the combustion of a fuel jet, delivered through a hole in the wall into a transverse airflow, has revealed the formation of spiral vortices in the flame of a jet (**Figure 10**).

It should be noted that the structure of the flame will differ for hydrogen and hydrocarbon fuel. In addition, it is known that the initial temperature of the fuel mixture can significantly change the reacting flow characteristics. For example, the diffusion flame of preheated

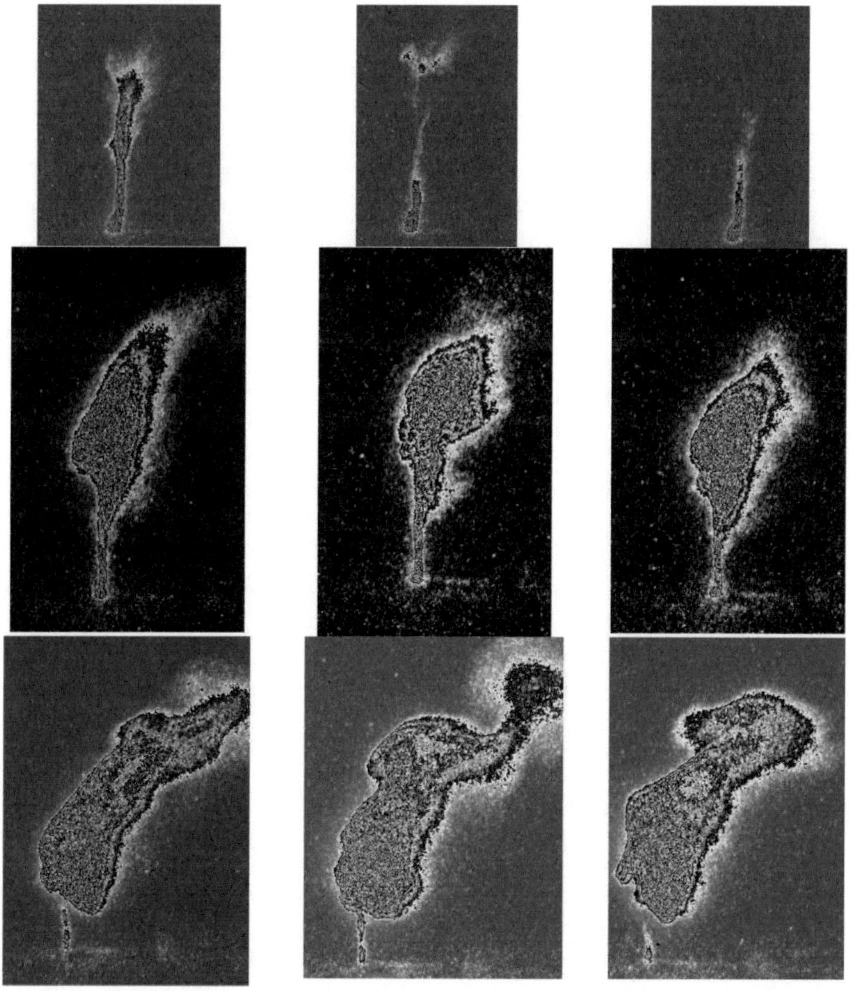

Figure 11. Chemiluminescence of OH radical in the wavelength range 306–308 nm with H_2/CO_2 jet combustion in air cross flows: d = 2 mm, U_0 = 60 m/s U_C = 8 m/s. Run 1, Y = 46%; Run 2, Y = 60%; and Run 3, Y = 70%.

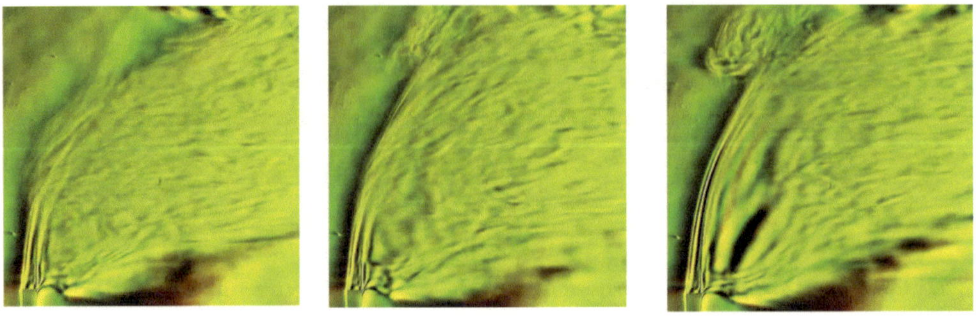

Figure 12. H_2/CO_2 flame jet in cross flow: d = 3.2 mm Re = 3476, Y = 39.4%, U_0 = 1.5 m/s.

propane is known to acquire the characteristic features of a hydrogen flame, in particular the formation of a thin luminous "filament" that connects the main flame body to the outlet of the fuel supply tube.

Photographing the chemiluminescence of the OH radical in the wavelength range 306–308 nm using the DiCAM-PRO ultraviolet camera is shown in **Figure 11**. The presented pictures were made in different, unrelated moments of time. Exposure time was chosen to provide the necessary level of illumination of the photosensitive matrix. In pre-extinction modes with maximum dilution of hydrogen by an inert gas (in mode Run 1), a torch body is not formed.

The flame-off at low hydrogen content (Run 1) achieved by gradual increase of the main flow velocity resembles the disappearance of a thin glowing filament. Under Run 2, all parts of the torch are observed until the moment of extinction. In Run 3, there are gaps in the body of the glowing filament. The change in the flame shape can be considered as a sign of a change in the mechanism of combustion stabilization.

New features are found for the diffusion flame in a cross flow at the instability regime in the jet source. As for combustion in still air, the intermittent character of the torch formation is revealed. In the laminar flow phase (**Figure 12** frame 360), an extended section of the longer-range jet is observed. When vortex structure appears, the flame topology changes dramatically: the flame length becomes shorter, and the curvature of the flame increases (frame 356). At a certain concentration of the diluent, the flame's body is divided into separate nonintersecting fragments.

7. Discussion and conclusion

In this chapter, hydrodynamics and combustion of a subsonic gas jet flowing out of the tube in the Reynolds number range 200–15,000 have been studied experimentally. The Reynolds numbers, characteristic of a laminar-turbulent transition in a pipe (both for cold-jet experiments and for combustion experiments), may differ due to the difference in the initial and boundary conditions in the pipe (velocity profile, turbulence level, non-isothermal conditions on the wall) [8]. However, such fundamental processes as intermittence and formation of vortex structures (puff) are present in the initial section both in the "cold" and in the "hot" jet.

The average velocity and velocity ripples in the near field of a propane-butane jet without combustion have been measured under isothermal conditions. Two "laminar" and "transitional" regimes have been established, in which it is possible to control mixing and combustion in the jet instability region at low Reynolds numbers (Re < 4000). In the "laminar" mode, when there is no puff, the instability of the jet flow prevails [2, 3], which allows controlling vortex structures, for example, by acoustic impact. For the "transitional" regime, the mechanism of two-stage instability is caused by the formation of turbulent spots inside the tube [7, 8] and the generation of vortex structures in the jet [2, 3]. The velocity of the puff structure is consistent with the convective velocity of the jet on the axis. However, the acoustic precursor spreads before the disturbance, which significantly changes the instability wave in the jet-mixing layer.

Several new results have been obtained in combustion experiments. Visualization and measurements have shown that the vortex structures (puff) that form in the transient flow-out regime have a strong effect on the flame. First, the vortex structures affect the instability of the outer flame front in different ways: for the mixture (C_3H_8/CO_2, Y = 46%), low-frequency oscillations (F = 6–15 Hz) remain (see **Figure 6**), and for combustion (H_2/CO_2, Y = 90%), the phenomenon of flickering can be suppressed almost completely (see **Figure 8**). Secondly, a reduction in the fuel content for a fixed Re number can lead to a transition from the attached flame to the disconnected flame. Thus, when mixtures of C_3H_8 with CO_2 or He burn in a transient mode, the flame is detached from the edge of the tube. Thirdly, depending on the set of parameters Re and Y, both a disconnected laminar flame and a disconnected turbulent flame were observed. Under such conditions, a transition from turbulent combustion to laminar combustion and vice versa is possible. Fourthly, there is a range of parameters depending on d, Re, and Y, at which the disconnected turbulent flame can be disrupted. Our data show that extinction under such conditions is associated with the action of turbulent spots arising in the transient flow regime in the tube, on the region of jet disintegration (see **Figure 7**). In the case of the formation of a flame in a drifting stream, the intermittent nature of the flame is observed in the puff formation mode. Its length increases in the laminar flow phase and decreases in the turbulent phase (see **Figure 11**).

In addition, previously unknown data on the topology of the heat-release regions of the diffusion flare have been obtained. Thus, when H_2/CO_2 burns on the outer boundary of the attached flame (without a drift), the passage of puff can be accompanied by the formation of "holes" in the flame, i.e., dark areas in which combustion apparently does not occur (see **Figure 8**). The visualization of the combustion of a H_2/CO_2 fuel jet, fed through a hole in the wall into a transverse airflow, has revealed the formation of spiral vortices in the flame of the jet (see **Figure 8**). A characteristic feature of a hydrogen flame in a transverse flow is the formation of a thin luminous "filament" that connects the main body of the flame with the outlet of the fuel supply tube (see **Figure 10**).

The influence of the vortex structure on two very different classes of jet flames (free jet and cross flow jet) is considered. A comparative analysis of these two flows allows to show the scale of the effect on the flow prehistory in the supply tube.

Along with the traditional methods of controlling the processes of mixing and combustion, an approach using the instability development regimes for controlling diffusion combustion, both in the jet stream itself and inside the source of jet formation, seems promising.

Acknowledgements

This work was partially supported by the Russian Foundation for Basic Research (Grant No. 17-08-00958). Hot-wire anemometer measurements were funded by the Ministry of Science and Higher Education of the Russian Federation.

Nomenclature

U_0	mean velocity on the axis in the initial section of the jet (m/s)
U_m	bulk velocity (m/s)
U_x	mean velocity on the axis of the jet (m/s)
u	root-mean-square value of velocity pulsations (m/s)
Y	volumetric concentration of fuel (%)
d	tube diameter (mm)
L	length of turbulent spot (mm)
l	tube length (mm)
x	longitudinal coordinate counted from the beginning of the jet (m)
$Tu = u/U_0*100$	turbulence degree of flow (%)
Tu_0	initial level of flow pulsation decrease (%)
$Re = U_m d/\nu$	Reynolds number of the jet
ν	kinematic viscosity of gas (m^2/s)
τM	time (s)

Author details

Dubnishchev Yuri Nikolaevich*, Lemanov Vadim Vladimirovich,
Lukashov Vladimir Vladimirovich, Arbuzov Vitali Anisiforovich and
Sharov Konstantin Aleksandrovich

*Address all correspondence to: dubnistchev@itp.nsc.ru

Kutateladze Institute of Thermophysics, Novosibirsk, Russia

References

[1] Abramovich GN. Theory of Turbulent Jets. Cambridge, MA: MIT Press; 1963. 671 p

[2] Ho CM, Huerre P. Mint: Perturbed free shear layers. Annual Review of Fluid Mechanics. 1984;**16**:365-424. DOI: 10.1146/annurev.fl.16.010184.002053

[3] Michalke A. Mint: Survey on jet instability theory. Progress in Aerospace Sciences. 1984; **21**:159-199. DOI: 10.1016/0376-0421(84)90005-8

[4] Peters N. Turbulent Combustion. Cambridge: Cambridge University Press; 2000. 303 p. DOI: 10.1017/CBO9780511612701

[5] Takeno T, Kotani Y. Mint: Transition and structure of turbulent jet diffusion flames. Progress in Astronautics and Aeronautics. 1978;**58**:19-35. DOI: 10.1016/S0082-0784(06) 80744-7

[6] Takahashi F, Mizomoto M, Ikai S. Mint: Transition from laminar to turbulent free jet diffusion flame. Combustion and Flame. 1982;**48**:85-95. DOI: 10.1016/0010-2180(82) 90117-1

[7] Lemanov VV, Lukashov VV, Abdrakhmanov R, Arbuzov VA, Dubnishchev YN, Sharov KA. Mint: Regimes of unsteady exhaustion and diffusion combustion of a hydrocarbon fuel jet. Combustion, Explosion, and Shock Waves. 2018;**54**:121-127

[8] Mullin T. Mint: Experimental studies of transition to turbulence in a pipe. Annual Review of Fluid Mechanics. 2011;**42**:1-24. DOI: 10.1146/annurev-fluid-122109-160652

[9] Roquemore WM, Chen LD, Seaba JP, Tschen PS, Goss LP, Trump DD. Mint: Jet diffusion flame transition to turbulence. Physics of Fluids. 1987;**30**:2600-2600. DOI: 10.1063/ 1.4738813

[10] Katta VR, Goss LP, Roquemore WM. Mint: Numerical investigations of transitional H_2/N_2 jet diffusion flames. AIAA Journal. 1994;**32**:84-94. DOI: 10.2514/3.11954

[11] Smits A, Lim T. Flow Visualization. Techniques and Examples. London: Imperial College Press; 2000. 306 p

[12] Kleine H. Mint: Schlieren imaging and the real world. Journal of Visualization. 2013;**16**: 193-199. DOI: 10.1007/s12650-013-0169-y

[13] Dubnishchev Yu N, Rinkevichyus BS. Methods of the Laser Doppler Anemometry. Moscow: Nauka; 1982. [in Russian]

[14] Dubnishchev YN, Chugui YV, Kompenhans Y. Mint: Laser Doppler visualization of the velocity field with elimination of the influence of multiparticle scattering. Quantum Electronincs. 2009;**39**:962-966. DOI: 10.1070/QE2009v039n10ABEH014054

[15] Dubnishchev YN. Mint: Laser Doppler visualization of the fields of three-dimensional velocity vectors with the use of the minimum number of CCD cameras. Quantum Electronics. 2010;**40**:551-555. DOI: 10.1070/QE2010v040n06ABEH014296

[16] Dulin VM, Markovich DM, Tolkarev MP, Chikishev LM. Mint: Application of modern optical methods for detecting the spatial structure of turbulent flames. Optoelectronics, Instrumentation and Data Processing. 2012;**48**(3):235-243. DOI: 10.3103/S8756699012 03003X

[17] Raffel M, Willert CT, Wereley ST, Yu K. Particle Image Velocimetry. A Practical Guide. Berlin: Springer; 2007. 448 p

[18] Arbuzov VA, Arbuzov EV, Berdnikov VS, Bufetov NS, Dubnishchev YN, Shlapakova EO. Mint: Optical diagnostics of the structure and evolution lets in a high-viscosity fluid. Optoelectronics, Instrumentation and Data Processing. 2014;**50**(5):466-473. DOI: 103103/ S8756699014050045

[19] Arbuzov VA, Arbuzov EV, Dvornikov NA, Dubnishchev YN, Nechaev VG, Shlapakova EO. Mint: Optical diagnostics of vortex ring–flame interaction. Optoelectronics, Instrumentation and Data Processing. 2016;**52**(2):66-72. DOI: 19.3103/S8756699016020084

[20] Arbuzov VA, Dvornikov NA, Dubnishchev YN, Nechaev VG, Novosyolova ON, Shlapakova EO. Mint: Hilbert diagnostics of vortex rings indused in air by a pressure pulse on a hole. International Journal of Spray and Spray and Combustion Dynamics. 2016, 2016;**8**(3):197-204. DOI: 10.1177/1756827716651763

[21] Arbuzov VA, Arbuzov EV, Dubnishchev Yu N, Sotnicov VV, Shibaev AA. Mint: Measurement of the velocity of Hilbert-visualized phase structures by the method of emulation of two-dimensional spatial filtering of their images. Optoelectronics, Instrumentation and Data Processing. 2016;**52**(2):66-72. DOI: 10.3103/S875669901606011X

[22] Reid RC, Prausnitz JM, Sherwood TK. The Properties of Gases and Liquids. New York: McGraw-Hill; 1977. p. 688

[23] Lemanov VV, Terekhov VI, Sharov KA, Shumeiko AA. Mint: An experimental study of submerged jets at low Reynolds numbers. Technical Physics Letters. 2013;**39**:421-423. DOI: 10.1134/S1063785013050064

[24] Aniskin VM, Lemanov VV, Maslov NA, Mukhin KA, Terekhov VI, Sharov KA. Mint: An experimental study of the flow of subsonic flat mini and micro air jets. Technical Physics Letters. 2015;**41**:46-49. DOI: 10.1134/S1063785015010034

[25] Lemanov VV, Terekhov VI, Sharov KA. Mint: Investigation of the flow in free and impinging air micro- and macrojets. Springer Proceedings in Physics. 2016;**185**:29-35. DOI: 10. 1007/978-3-319-30602-5_4

[26] Hallberg MP, Strykowski PJ. Mint: On the universality of global modes in low-density axisymmetric jets. Journal of Fluid Mechanics. 2006;**569**:493-507. DOI: 10.1017/S002211200 6002357

[27] Nikitin NV, Pimanov VO. Mint: Numerical study of localized turbulent structures in a pipe. Fluid Dynamics. 2015;**50**:655-664. DOI: 10.1134/S0015462815050075

[28] Nikitin NV, Pimanov VO. Mint: Sustainment of oscillations in localized turbulent structures in pipes. Fluid Dynamics. 2018;**53**:65-73. DOI: 10.1134/S0015462818010111

[29] Lyons KM, Watson KA, Carter CD, Donbar JM. Mint: On flame holes and local extinction in lifted-jet diffusion flames. Combustion and Flame. 2005;**142**(3):308-313. DOI: 10.1016/j.combustflame.2005.04.006

[30] Steinberg AM, Sadanandan R, Dem C, Kutne P, Meier W. Mint: Structure and stabilization of hydrogen jet flames in cross-flows. Proceedings of the Combustion Institute. 2013;**34**: 1499-1507. DOI: 10.1016/j.proci.2012.06.026

[31] Sullivan R, Wilde D, David RDR, Seitzman JM, Lieuwen TC. Mint: Time-averaged characteristics of a reacting fuel jet in vitiated cross-flow. Combustion and Flame. 2014;**161**: 1792-1803. DOI: 10.1016/j.combustflame.2013.12.022

[32] Wu Y, Lu Y, Al-Rahbi IS, Kalghatgi GT. Mint: Prediction of the liftoff, blowout and blowoff stability limits for pure hydrogen and hydrogen/hydrocarbon mixture jet flames. International Journal of Hydrogen Energy. 2009;**34**:5940-5945. DOI: 10.1016/j.ijhydene.2009.01.084

[33] Menon R, Gollahalli SR. Mint: Combustion characteristics of interacting multiple jets in cross flow. Combustion Science and Technology. 1988;**60**:375-389. DOI: 10.1080/00102208808923994

[34] Kalghatgi GT. Mint: Blow-out stability of gaseous jet diffusion flames. Part I: In still air. Combustion Science and Technology. 1981;**26**:233-239. DOI: 10.1080/00102208108946964

[35] Kalghatgi GT. Mint: Blow-out stability of gaseous jet diffusion flames. Part II: Effect of cross wind. Combustion Science and Technology. 1981;**26**:241-244. DOI: 10.1080/00102208108946965

[36] Dahm WJA, Maymant AG. Mint: Blowout limits of turbulent jet diffusion flames for arbitrary source conditions. AIAA Journal. 1990;**28**:1157-1162. DOI: 10.2514/3.25186

[37] Moore NJ, McCraw JL, Lyons KM. Mint: Observations on jet-flame blowout. International Journal of Reacting Systems. 2008;**2008**:461059. DOI: 10.1155/2008/461059

Chapter 4

Vortex Generation, Experimental Characterization, and Application in Turbulent Flows

Guillermo Martin Capittini,
Julio Marañon Di Leo and Juan Sebastian Delnero

Additional information is available at the end of the chapter

http://dx.doi.org/10.5772/intechopen.79504

Abstract

This chapter is focused on vortex detection, generation, and characterization. There are many ways to generate and characterize vortex; this chapter is focused on two techniques to generate a vortex, with fixed surface, often called vortex generators (VG), and with blowing air. Vortex detection is one of the major problems in fluid dynamics and usually some characteristics of the vortex must be known in order to detect them; once, the vortex is detected, a velocity analysis is helpful to characterize it. Most characterization techniques involves the comparison of some known properties of vortex, such as velocity field, vorticity field or tensor, turbulence intensity, etc. The technique to be used to characterize a vortex is closely related to the data that one possesses. In measuring methods such as particle image velocimetry (PIV), there are algorithms that can easily detect size and vortex centers, relaying in velocity and vorticity. This chapter focuses on detection by analyzing velocity signals, via wavelet transform and statistical properties. When it is not possible to characterize a vortex because it does not have a coherent structure, another approach must be used such as defining turbulence intensities and zone of influence of the vorticose structure.

Keywords: vortex, detection, characterization, turbulence, generation

1. Introduction

Determining exactly how a system is going to behave in aerodynamics can be a challenge. This is because the incident flow plays a main role in the behavior. A wing immersed in a turbulent flow is not going to experiment the same forces as it would in a laminar flow. Even a change in the turbulent scales can conduct to a variation in the aerodynamic coefficients of the wing. This phenomenon can become much critic for large vortex, where even the relative position of the vortex over the wing is going to play a role in the wing efficiency.

For decades, this reason has led scientist and engineers to study systems' behavior in different conditions of the free flow. To do this, one must be able to change the flow conditions in terms of turbulence and turbulent scales, here is where different kinds of vortex and turbulent generators came along. There is a wide variety of techniques to create vortexes; an usual and widely used is Von Karman streets as in [1], or delta wing VG as in [2–4]. These methods involve fixed surfaces, but there are others where energy and mass are injected in the flow, to create the vorticose structure. Once, the vortex is created, a full characterization must be performed to determine the flow condition. In modern measurement techniques such as particle image velocimetry (PIV), very effective algorithms can be used to characterize the flow, such as [5], but for anemometric measurement, this can represent quite a challenge.

This chapter's intention is to give a good introduction to vortex identification techniques, as well as the characterization and the generation, and in the ending showing results of how the presence of different vorticose flow configurations can lead to a modification in a system behavior.

2. Characterization

There are different ways to characterize a vortex. As told before, this chapter covers the characterization of vortex measured with anemometers. The idea here is to find if there is a vortex passing by analyzing the evolution of velocity over time in a given point or in set of points. This section mainly focuses in the use of continuous wavelets transform to identify a vortex immersed in a turbulent flow.

Let us consider for porpoise of demonstration a simulated signal. To do so, lets define the velocity distribution of a 2D vortex. There are many kinds of vortex, rotational and irrational, physically plausible and not, and for this example and for the purpose of demonstration, a Rankine vortex is going to be used. This kind of vortex is an approximation of a real one, since it does not take into account the effect of the fluid viscosity. Due to the neglect of viscosity, these kinds of vortexes do not vanish over time. Another more general one is a Lamb-Oseen vortex, this is an exact solution to the Navier–Stokes equations, and it does take into account the effect of viscosity. But for the purpose of this simulation, there is no need to use it.

A Rankine vortex has a velocity distribution as shown in Eq. (1) as shown in [6].

$$V_\theta(r) = \begin{cases} \dfrac{\Gamma r}{2\pi R^2} r < R \\ \dfrac{\Gamma}{2\pi r} r < R \end{cases} \tag{1}$$

It is seen in this distribution that the flow velocity increases linearly until it reaches a radius R and then decreases like 1/r as shown in **Figure 1**.

http://dx.doi.org/10.5772/intechopen.79504

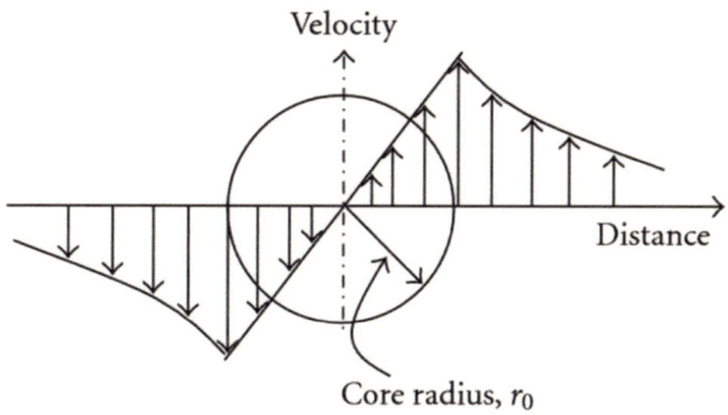

Figure 1. Velocity distribution on a Rankine vortex.

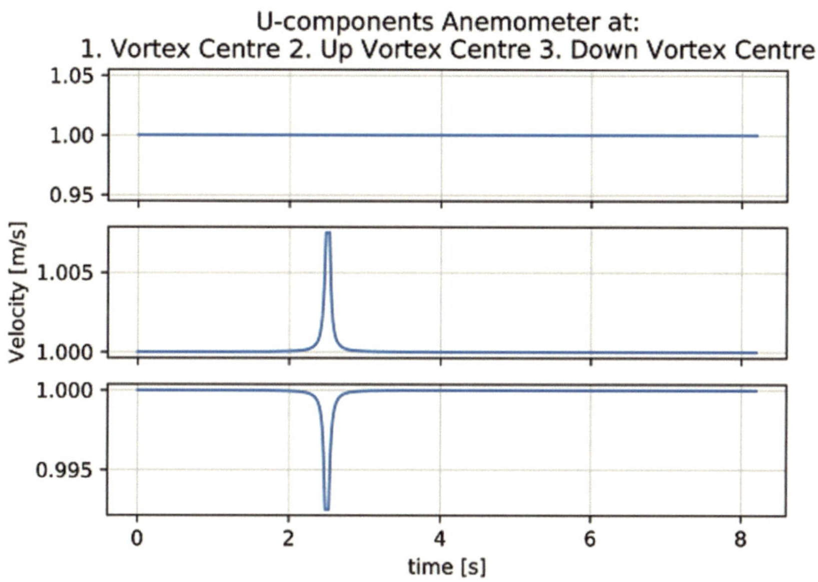

Figure 2. Velocity distribution on a Rankine vortex U-component.

Now, if a single vortex is moving with flow velocity, and an arrangement of anemometers is settled to capture the U-component (flow direction) and V-component (perpendicular to flow direction), in case the anemometer captures the vortex, there are three possibilities, to capture it exactly on its center, up, or down to it. **Figures 2** and **3** show the signals that would be obtained in such cases.

As it is shown, the V-component signal shape does not change substantially, but the U-component, clearly evidence if the vortex is over or under the anemometer, for a clockwise vortex, it would be opposite for a counterclockwise. Because of the velocity distribution of a vortex, the U-component is not going to evidence by itself the passage over an anemometer.

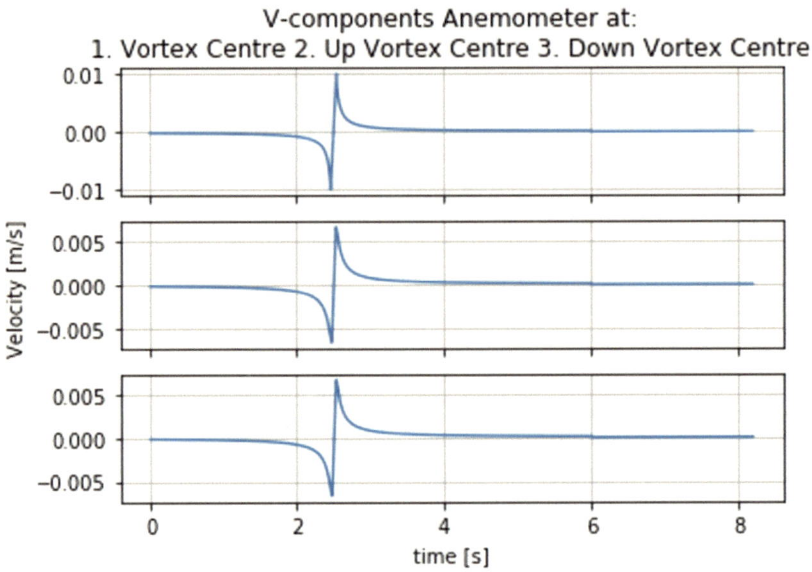

Figure 3. Velocity distribution on a Rankine vortex V-component.

Watching the V-component, one can clearly distinguish that a vortex has passed through the anemometer, by the change in the velocity sign.

This simplified case shows our goal, that is, to determine if this kind of structure is present in a velocity signal.

In these simple cases, a vortex passage is easy to identify, but in the presence of a highly turbulent flow, this task is not so trivial, and more advance techniques or experimental setups are required.

For simplicity, from now on, only an anemometer above the vortex center is going to be considered. It must be said that in an experimental setup, one usually do not know where exactly the vortex center is going to be, so multiple measurements must be taken until this position is determined. In a steady flow, a coherent structure becomes periodic, so the passage would repeat over a time t_p, and a signal similar to the one shown in **Figure 4** is expected.

Given this type of signal, a useful tool in vortex detection can be introduced, that is, the continuous wavelet transform (cwt). The cwt is useful to find coherent structures of a certain shape inside a signal, for more information of detection using wavelet transform, see [7–9]. To detect these structures, there are two useful wavelets: the derivative of Gaussian function of order 1 (DOG1) and the one of order 2 (DOG2) (also known as the Ricker wavelet). Both wavelets are plotted in **Figure 5**. Using DOG1 for the V-component and DOG2 for the U-component, one is able to detect a vortex passage.

It is seen in **Figures 6** and **7** that every time a vortex goes through the anemometer, a maximum of the wavelets coefficient occurs.

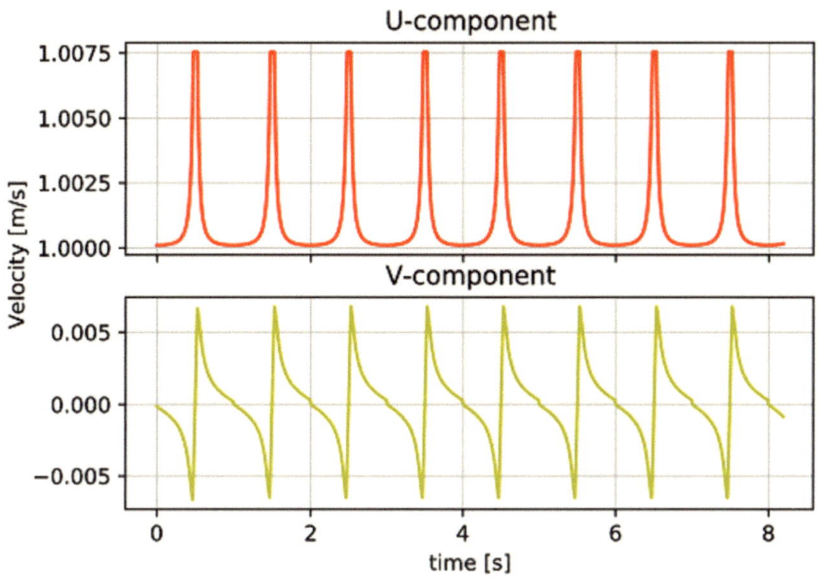

Figure 4. Velocity signal for a periodic vortex.

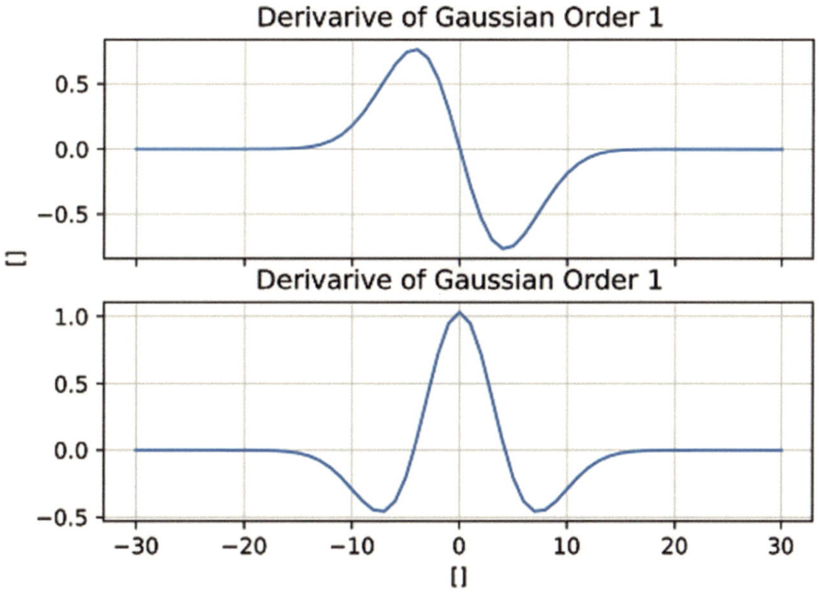

Figure 5. Wavelets used for detection.

This may seem trivial since one can easily detect the vortex passage in the velocity signals. But when turbulence is present in the free flow, it becomes harder to determine if and when a vortex is passing.

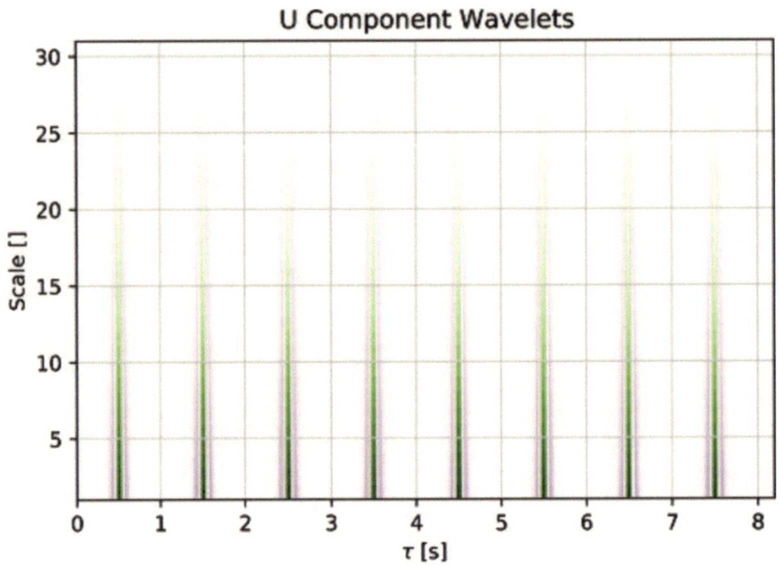

Figure 6. Wavelets coefficients in U-component.

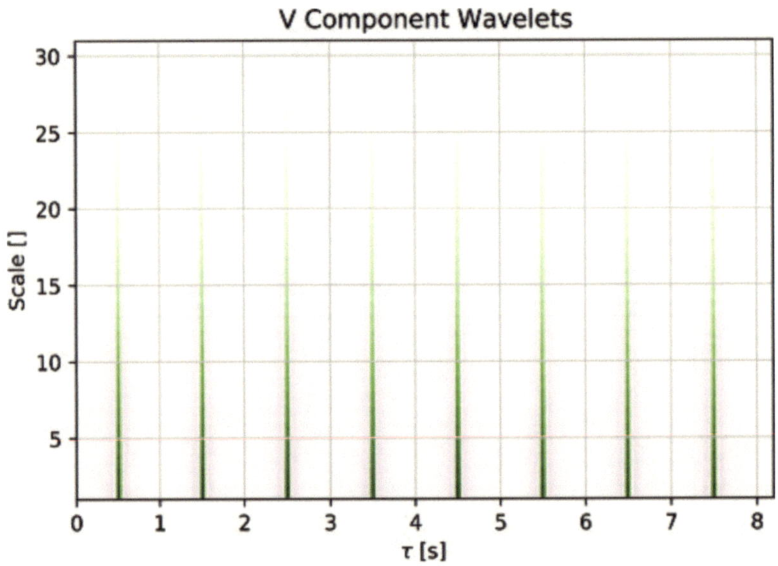

Figure 7. Wavelets coefficients in V-component.

To consider the mentioned problem, from now on, a turbulent flow is going to be treated. For this purpose, a random signal is added to our periodic vortex. This signal is going to simulate turbulence, and it is going to be generated by a random noise with a Gaussian distribution. It must be said that this is for demonstration purposes only, since a random noise does not fully represents a turbulent flow, and does not fulfill the constrains imposed by the "Kolmogorov – 5/3 spectrum."

But the techniques presented here, give similar results for these simulated signals with Gaussian noise and for measured signals with real turbulence.

Let us first define the so-called turbulence intensity. This property is useful to characterize a turbulent flow, since the vortex present in it is usually noncoherent, this property gives us, not the velocity function, but a measure of how much energy is present in the turbulence. And it is defined as in Eq. (2)

$$T_i = \frac{\sigma_u}{\mu_u} \quad (2)$$

where σ_u represents the standard deviation of the velocity component, and μ_u the mean value of the same component.

For this simulation, a turbulence intensity of 2% is going to be set; this is a normal value for wind tunnel testing and the maximum value of the vortex tangential velocity is going to be 3% of the flow mean velocity, and with a characteristic radius of 4 cm. **Figures 8** and **9** shows both components of velocity and a wavelet map for the V-Component.

It can be seen that when turbulence is present in the free flow, a vortex passage is not easily found. But **Figure 9** shows that analyzing the wavelet coefficients of the DOG1 in the V-component, the passage becomes obvious. This becomes a powerful tool, since it allows one to detect a vortex passage even when the turbulent velocity fluctuation is in the order of the maximum vortex velocity.

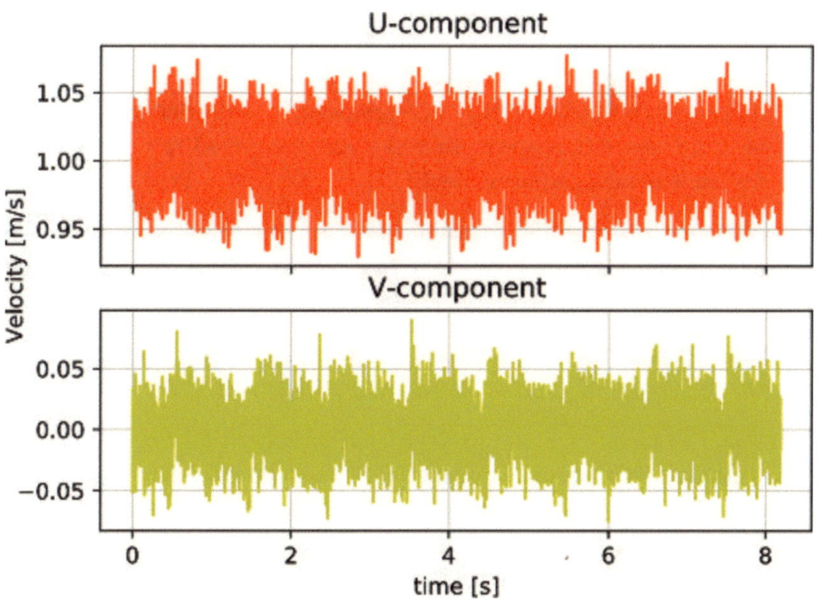

Figure 8. Velocity signals for a periodic passage with turbulence.

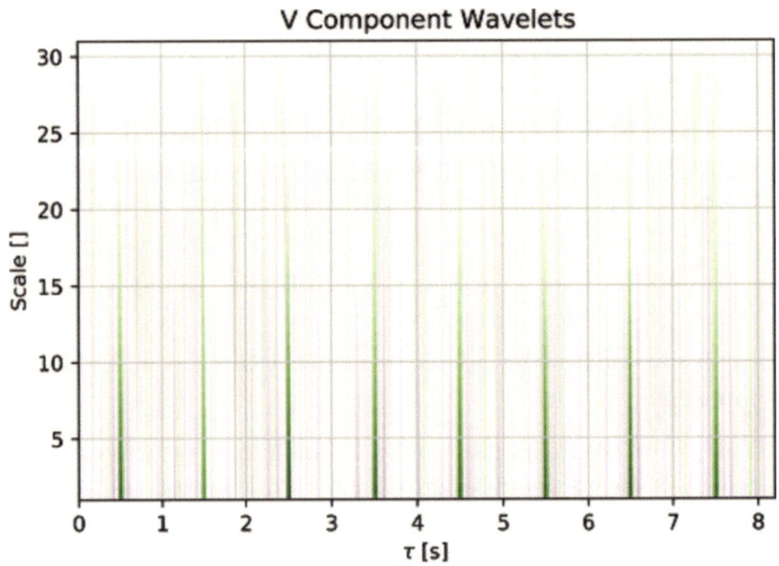

Figure 9. cwt on the V-component signal, with turbulent free flow.

The last technique to be presented here is how to detect a vortex passage when the turbulent velocity is higher than the vortex velocity. This is not always possible, but it can be done in some cases. The way to accomplish this is by taking more than one measurement synchronize with the vortex passage or with some phenomena that generate the vortex (this could be an oscillating surface, a pulsing jet, or the beginning of a movement that creates an steady state) and then take the mean of the measurements. By doing this, the turbulence present in the signals is going to be attenuated, and the periodic phenomenon present in all the signals is going to remain. After a wavelet transform can be used to detect the vortex passage, it must be said that if a complete synchronization is not achieved, a distortion may be present in the remaining signal, but if the difference in time is small compared to the time, it takes to the vortex to pass, the same results can be achieved.

Figures 10 and **11** show the results accomplished with the same simulation as in the case before but with a vortex velocity of 0.5% of the free flow velocity, using 15 different measurements.

It can be seen that, when the mean of different measurements the cwt shows the presence of the vortex.

This approach taken so far allows one to detect the presence of a vortex. To characterize it, there are many different techniques. Once the vortex is detected with the cwt, the maximum coefficient gives us the time (x axis) it passed and the scale (y axis) that it has respect to the wavelet scale. This scale is related to the time (or pseudo frequency) it took to the vortex to pass the point for further readings see [10]. The frequency shown in Ref. [10] relates to the time the vortex took to pass the anemometer as $1/f$, where f is the wavelet pseudo frequency, by knowing the mean velocity at the point the vortex size can be obtained with Eq. (3).

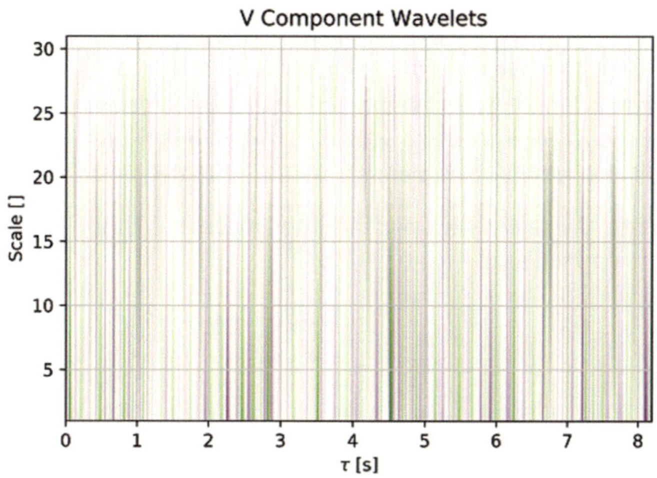

Figure 10. cwt on the V-component signal, with turbulent free flow, 1 measurement.

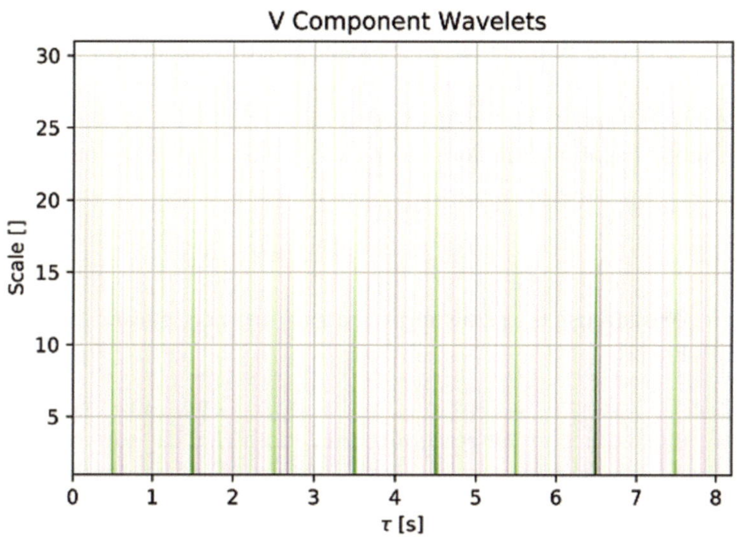

Figure 11. cwt on the V-component signal, with turbulent free flow, mean of 15 measurements.

$$S_v = V*t \tag{3}$$

S_v, is the vortex length or size, V is the point mean velocity, and t is the inverse of the pseudo frequency.

Each type of vortex would need a particular characterization, since it depends on the velocity distribution it has. For Rankine vortexes, the radius (or length) and the circulation Γ are sufficient. This last one is harder to obtain in the presence of a turbulent flow, but with a clean signal, different measurements points can be taken in height and relating the peak velocity of them to the velocity distribution in Eq. (1). For any other vortex, the velocity distribution must be known.

This is a helpful approach and gives one the tools in case the vortex has a known form. But in presence of multiple vortexes interacting with each other, this is not the best suited approach. For these kinds of flows, another parameters must be presented, such as representative scale.

To obtain a representative scale of the flow, the autocorrelation of the signal must be calculated first. This function is defined as in Eq. (4)

$$R(\tau) = \frac{\int u(t)*u(t-\tau)d\tau}{\int u(t)*u(t)dt} \tag{4}$$

As defined in Eq. (4), the autocorrelation function has a maximum value in $\tau = 0$ and is: $R(0) = 1$. This function is obviously discrete for measured signals and one only know the values for the sampled times. To define the representative scale, different criteria can be adopted, but most of them give similar results. Using them, one is able to obtain a representative scale of the smallest and the biggest vortex present in the flow, these are the microscale and macroscale, respectively. Here, three criteria are going to be introduced as presented in [11], one for the macroscale and two for the microscale. For the macroscale, the most commonly used criterion is the 0-passage. One interpretation for this criterion is that as long as the correlation does not become 0, the biggest vortex in the flow field is passing, once the vortex has passed, the autocorrelation function becomes 0. So, the 0-passage criteria would be as in Eq. (5).

$$\tau \textit{for the first } R(\tau) = 0 \tag{5}$$

After obtaining τ, one can use the same approach as with wavelets and define the macroscale of the vortex with Eq. (3).

For the microscale, the two criteria used are the "fist point slope," and the $\mathbf{1}/\boldsymbol{e}$. This criteria are far less intuitive and require much reasoning and mathematical support. This exceeds the purpose of this chapter and there are not going to developed here, but for further reading see [6].

The first point slope criterion states that the microscale (scale in time) present in a turbulent flow, or in a flow with multiple vortexes interacting, is the time obtained by extrapolating a straight line that has the point (0,1) (first point of the autocorrelation function) and a slope equal to the slope of the autocorrelation function in its first point. This is as stated in Eq. (6).

$$\tau = -R(0)/R'(0) \tag{6}$$

As stated before, $\boldsymbol{R}(\mathbf{0})$ is the maximum value of $\boldsymbol{R}(\boldsymbol{\tau})$ so $\boldsymbol{R}'(\mathbf{0})$ must be negative, this gives a positive $\boldsymbol{\tau}$. As before, Eq. (3) can be used to compute the size of the vortex.

The last criterion is the $\mathbf{1}/\boldsymbol{e}$. This criterion simply states that the value of the time microscale is equal to the value $\boldsymbol{\tau}$ for which $\boldsymbol{R}(\boldsymbol{\tau})=\mathbf{1}/\boldsymbol{e}$ as in Eq. (7).

$$\tau \; for \; R(\tau) = 1/e \tag{7}$$

This gives one a characterization of the spatial scales. Another useful approach is to determine the power spectrum density of the signal; this tells us how the energy is distributed in frequency. And can be useful in case a periodic phenomenon is present on the flow field.

The previous are the more classical techniques. But many more can be used. In general, to characterize any flow, one needs to have some knowledge of the phenomenon to capture. In the next section, other methods are going to be shown, for example, how an array of turfs can be useful to determine a longitudinal vortex diameter. Cross-correlation between spatial signals can help to determine if a certain flow structure is propagating or if it is been defused by turbulence, with an spatial anemometry array or PIV one can find the vortices field by taking the spatial derivatives of the velocities, and many for techniques are available, for further reading see [12].

3. Vortex generation

In the previous section, the main goal was to show how to characterize a passing vortex, this can be useful when a phenomenon is produced in an experiment and one needs to understand it, but vortex are always used for many different purposes in aerodynamics, one of the most simple approaches to passive flow control are vortex generators. These fixed surfaces are able to generate vortexes or stochastic turbulence in some cases. That is used to modify the behavior of a system. Also, when working with aerodynamic comfort for civil structures, or even to find the load in such structures, is always needed to test them inside a wind tunnel. In order to have a realistic view of the problem, the turbulence present in the atmospheric boundary layer must be modeled inside the wind tunnel. In these cases, one needs to generate vortex of certain scales, and to use the techniques presented before in order to determine if the turbulence generated fulfill the modeling criteria.

In this section, some useful techniques to generate vortexes of certain scales are presented, and what variables must be taken into account to accomplish a desired behavior.

The first method to be presented is the generation of a longitudinal vortex. This method is very effective, and allows the generation of a vortex with a well-defined diameter. A longitudinal vortex is a rotating structure along a define axis, this vortex has the particularity of changing its diameter along the axis. One common example of it is the wing tip vortex, present on any airplane, **Figure 12** shows an illustration of the phenomena.

The vortex shown in **Figure 12** is a result of the perturbation a wing generates downwards the flow, and it is a result of the tridimensionality of the wing. In practice, the presence of these kinds of vortexes result in a potential risk to an airplane, since their presence can drastically change the wing behavior. For those reasons, the study of such vortexes is highly important.

A way of generating these kinds of vortexes is by injecting compressed air at a certain incidence angle inside a tube. The resulting structure has a diameter in the order of the tube's

Figure 12. Longitudinal vortex representation.

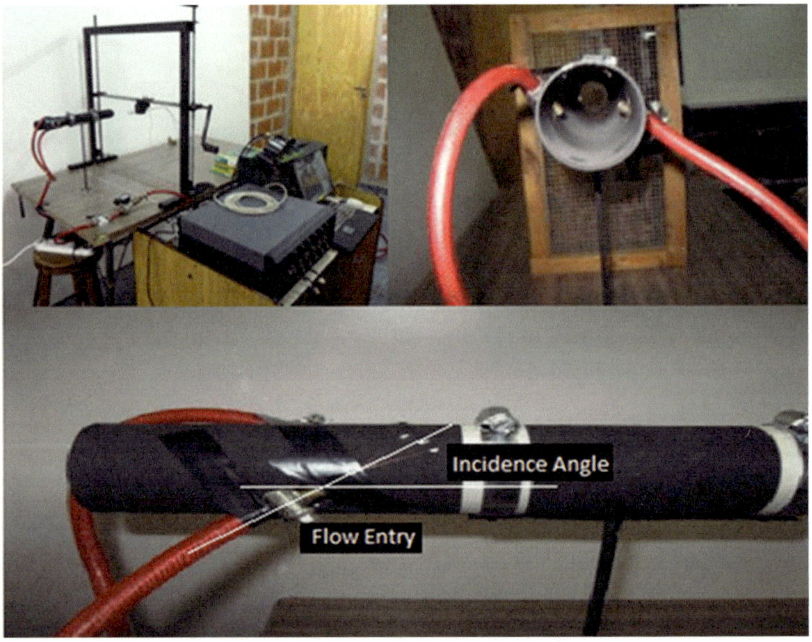

Figure 13. Experimental set-up for longitudinal vortex generation.

diameter, and the velocities are a function of the pressure, flow rate, and incidence angle of the injected air. An experimental setup of such array is presented in **Figure 13**. Inside the tube, two hoses are connected with a fixed angle respect to the tube transversal plane: Pleasure and flow regulators are set between the tube and the air compressor to fix pressure and flow rate to a known value. The injection of air at a certain angle respect to the cylinders axis generates a rotary field, since the air must follow the surface of the cylinder, when this rotatory flow merges with the free flow entering the tube, a helical flow pattern is produced, and when the flow leaves the tube a helical vortex is generated, and then diffused as it moves in the flow

http://dx.doi.org/10.5772/intechopen.79504

direction. To characterize the vortex, a hot wire anemometer is set in front of the tube and spatial measurements of the velocity are taken, precisely the U-component and the V-component.

In this case, that the vortex has a well-defined diameter, a tufts array is handy to determine the diameter evolution along the axis, at different longitudinal positions downstream the vortex generator. This tufts array consists of a steel wire mesh with a 1 cm spacing, where thin threads of adequate color and a 4 cm length, to provide a good flow visualization, were fixed at each vertices of the mesh. The size of the turf array was about 44.5 cm height and 24 cm width. It can be seen in **Figure 14** how the tufts rotate and approximately define the vortex diameter.

In this setup, the tube diameter determines the size of the vortex, and a higher pressure and flow rate decreases the widening. A suitable characterization for these kinds of vortexes is the vortex size along the axis and the velocity distribution along the radius. **Figure 15** shows the vortex size evolution and **Figure 16** the velocity distribution.

The blue curve in **Figure 16** represents the U-component of the velocity and the red one the V-component for a tube with 6 cm diameter, 30 cm long and 45° of the flow entry. The rotational effect becomes clear by looking at the V-component, it evidenced by the change in the velocity sign. It is seen in **Figure 16** how velocity changes with the radius, and how it becomes 0 near the vortex size. This means that this are concentrated vortexes that rapidly disappear. As it is expected when the vortex moves, its diameter grows and the velocity diminish, this is a consequence of viscosity diffusion.

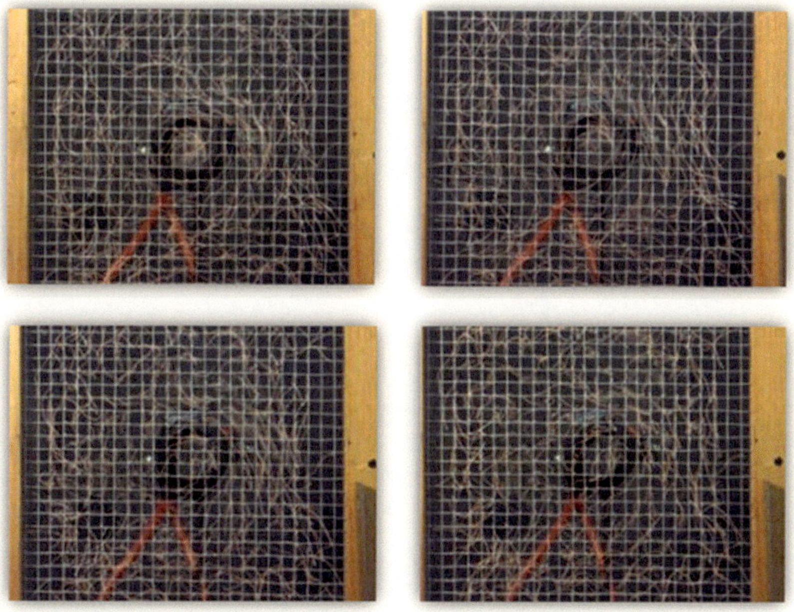

Figure 14. Visualization of the vortex diameter.

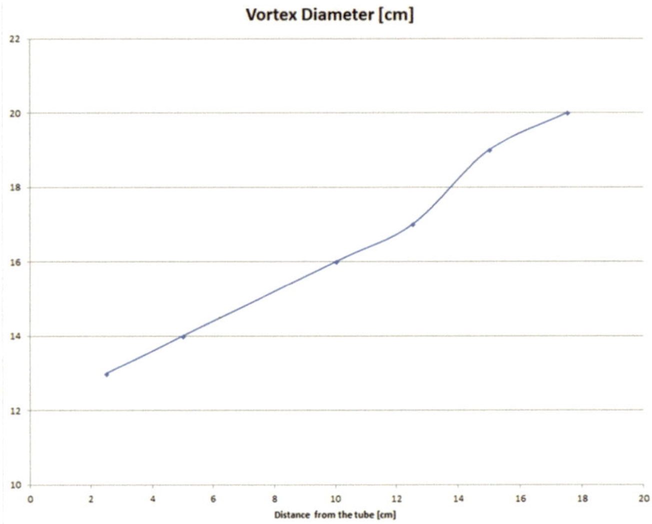

Figure 15. Vortex diameter for a 45° incidence and 4 bar pressure vs. distance from the vortex generator.

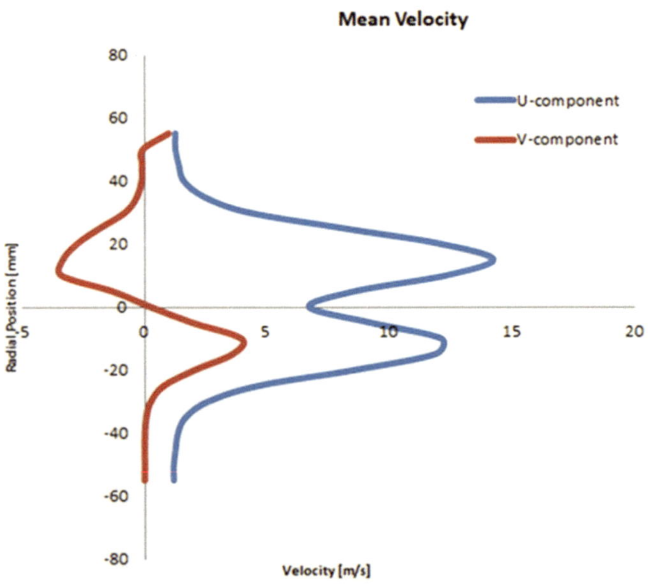

Figure 16. Velocity vs. radius. In red is the tangential component, in blue the axial component.

Another widely used method for generating vortex are fixed surface, in some cases called vortex generators (VG). This method is much simpler than the one presented before since it only requires a fixed surface. Almost any surface facing the flow will create vorticity or

Figure 17. Fin-type vortex generator, with a hot wire anemometer downwards.

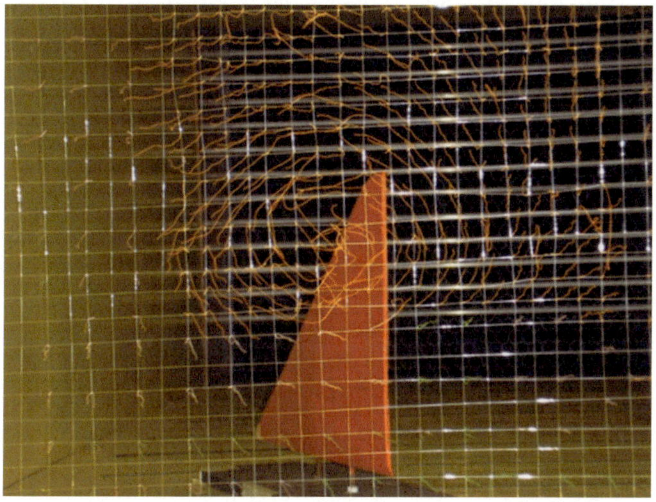

Figure 18. Fin-type vortex generator, producing a longitudinal vortex.

turbulence downwards itself. This method is usually used in airplane wings to delay the detachment of the boundary layer, this is possible by inserting energizes flow from the free flow to the boundary layer. **Figure 17** shows an example of them.

The main complexity with this method is that the generated vortex does not always have a defined structure. A delta wing or a fin, generates longitudinal vortex, but not in any Reynolds number or under any angle of attack. So the scales and the overall structure of the vortex can heavily depend among the flow condition. For this reason, these kind of devices are not usually characterized alone, but with the system they are meant to work with. **Figure 18** shows a longitudinal vortex generated by a fin VG, and a tufts array shown a helical vortex similar to the ones produced by a wing, the results of this experiment are going to be treated in the last section of the chapter.

When it is not possible to find a defined structure of the generated vortexes, the change in the scales, and turbulent intensity between the flow upstream the VG and downstream is a good way of quantify the VG behavior. Even if this approach does shows the full structure of the vortexes present in the flow, it allows one to estimate where the zone of influence is located; this is helpful to decide the separation of a VG array, for example, and allows one to know if the turbulence generated by one VG is going to interact with the one generated by its neighbor.

For fin-type VG, the main characteristic that will determine the turbulence intensity behind itself is the angle of attack or angle of attitude related to the flow. Increasing the angle of attack will increase the turbulence intensity behind itself, and will also increase the drag force over the VG, when vortex generation is the only concern, this is not a problem, but in applications like an airplane, this feature must be taken into account. More detailed information can be found in [13].

Figure 19 shows an experimental setup to determine the influence zone of a fin-type VG of 10 mm height and 25 mm long, positioned over a flat plate. In each of the planes, anemometric measurements were taken and the turbulence intensity present behind the VG was compared to the one of the free flow. For each, a point grid was generated to take the measurements, the planes shown are positioned at 12.5, 25, 50, and 100 mm, behind the VG. **Figure 20** shows the results of the turbulent intensity in the V-component, for the first and last plane shown in **Figure 19**, for a 6 m/s flow velocity, a 30° angle of incidence

It must be said that mean velocities in the point grid were analyzed too, but the results did not showed a coherent structure present in the flow field. But with the turbulence intensity field, one can see that the VG is able to change the overall structure of the flow. This property is extremely helpful when designing a flow control system, particularly a system whose intention is to trigger the turbulent transition of the boundary layer. The changes on the flow structure (as seen in **Figures 20** and **21**) are mostly held in the right side of the plane, this is due to the fact that the rotation of the VG (The 30° of incidence) left the tail positioned on the right side. The low pressure that forms in the back plane of the VG makes the flow to curl from the front plane to the back one, so it is expected that the velocities induced by the VG were stronger near the tail.

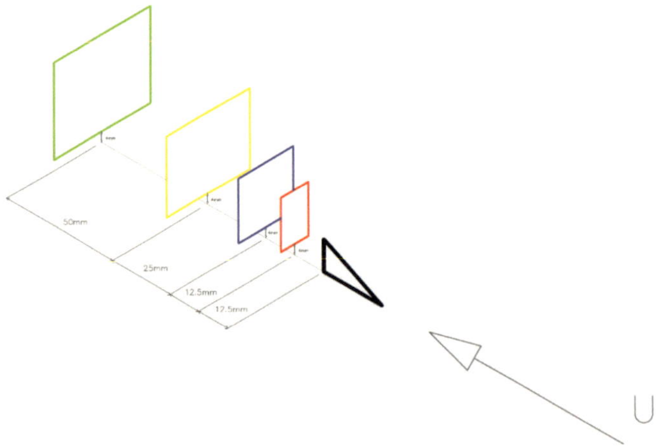

Figure 19. Grid planes for characterization of a fin-type vortex.

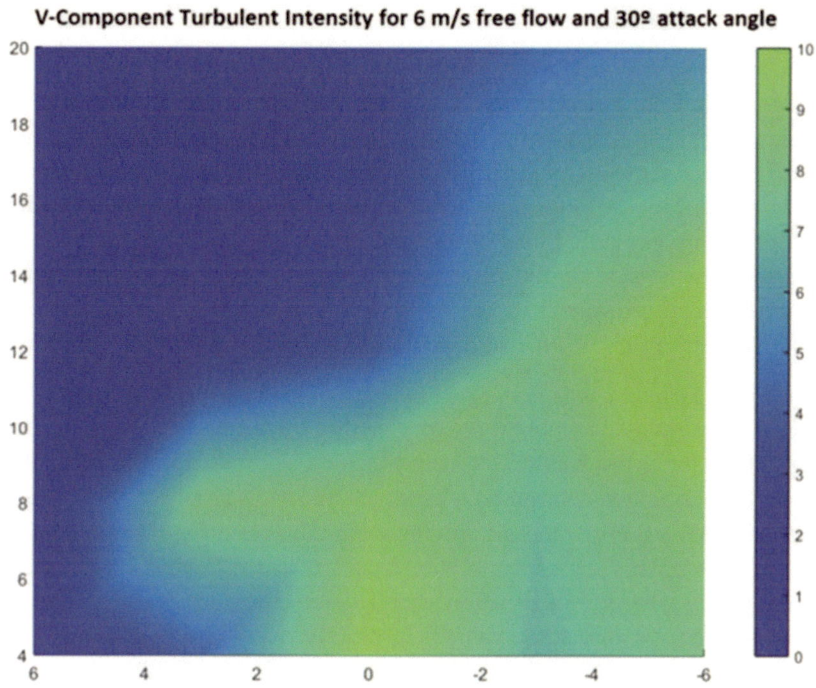

Figure 20. Turbulent intensity of the V-Component in the first plane, x and y axes are in mm, the VG is at position (0,0 in the plane).

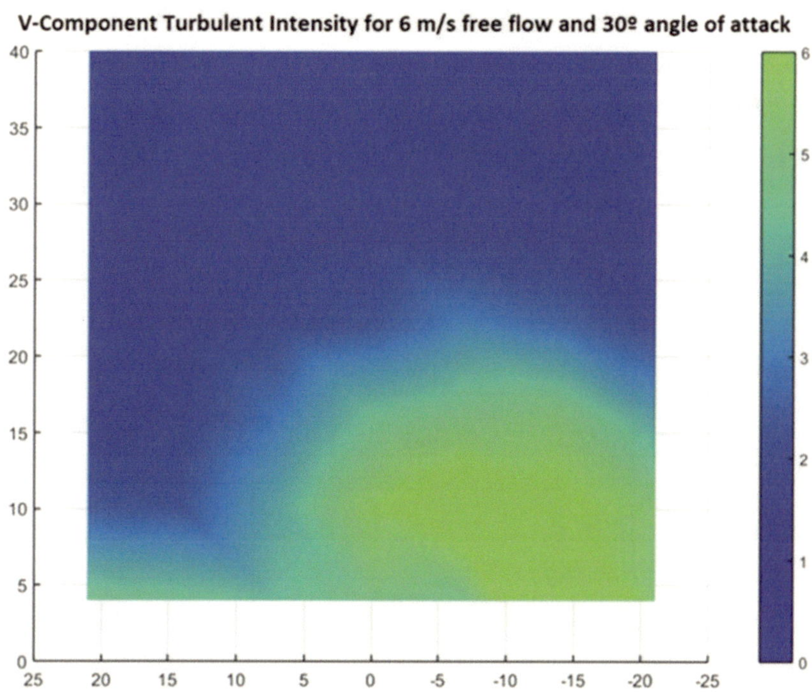

Figure 21. Turbulent intensity of the V-Component in the fourth plane, x and y axes are in mm, the VG is at position (0,0 in the plane).

4. Application example

This chapter shows how to detect, characterize, and generate vortexes; in order to give a full picture of their utility, in this last section, an application is shown. Vortexes are useful in flow control, needed to determine loads in civil structures, and can substantially change the behavior of a system when they are present in a free flow. To scratch the surface of this discipline, experimental results are presented here of the behavior of a wing in presence of a longitudinal vortex of a scale whit a size in the order of the wing chord. As it was mention before, longitudinal vortexes are produced by the wing tip of an airplane, so this approach can give us for example an insight of how an airplane it going to be affected in a formation flight.

The vortex in this experiment was generated by the fin VG presented in **Figure 18**. **Figure 22** shows a schematic of the experimental setup. For this experiment, a wing model was positioned in the wind tunnel test section, attached to force sensors, to determine the forces the wing produces. The model was placed between two end plates as shown in **Figure 22**, in order to minimize the 3D effects. Force measurements were taken and the Cl and Cd dimensionless coefficient determined. This experiment was repeated for different vortex configurations, which consisted of the same vortex positioned at different heights. One with the vortex center below the wing, one above, one at the same height of the wing chord, and one with no vortex (clean wing). As mentioned before, the vortex was generated with a VG, in this case, a coherent structure was produces, and its size was determined with a tufts array. Since the intention of this section is only to show how a system is modified when it is present in different vorticose structures, the full characterization of the vortex generated for this experiment is not relevant, and it is only going to be mentioned that the vortex size was similar to the wing chord.

Figures 23 and **24** show the C_l and the efficiently $E = C_l/C_d$ of the wing for the different configurations mentioned before.

Since no mass or momentum was added to generate the vortex here, the behavior of the wing in presence of the vortex, it is only a result of the coupling of the flow fields generated by both,

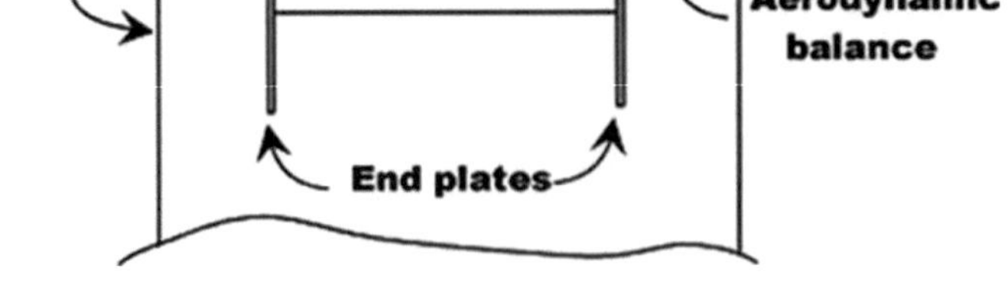

Figure 22. Experimental setup for the experiment, the fin is placed upwards the model.

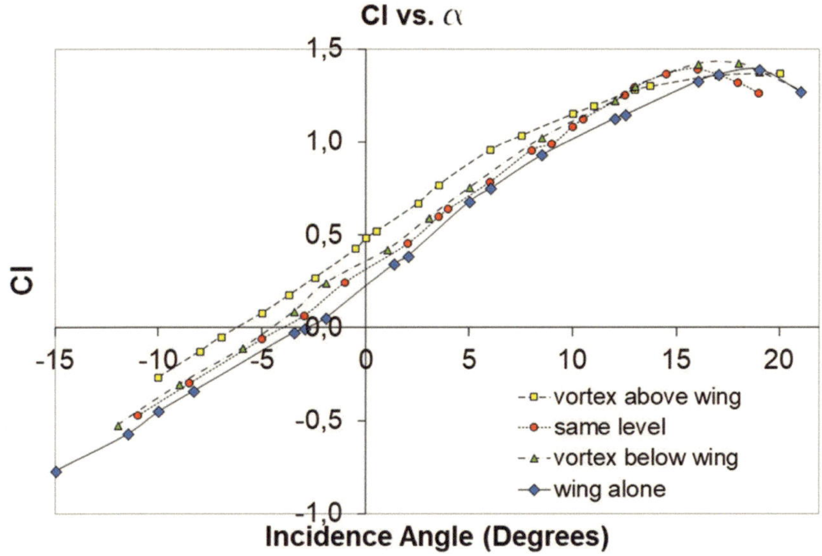

Figure 23. C_l coefficient for the wing in different conditions.

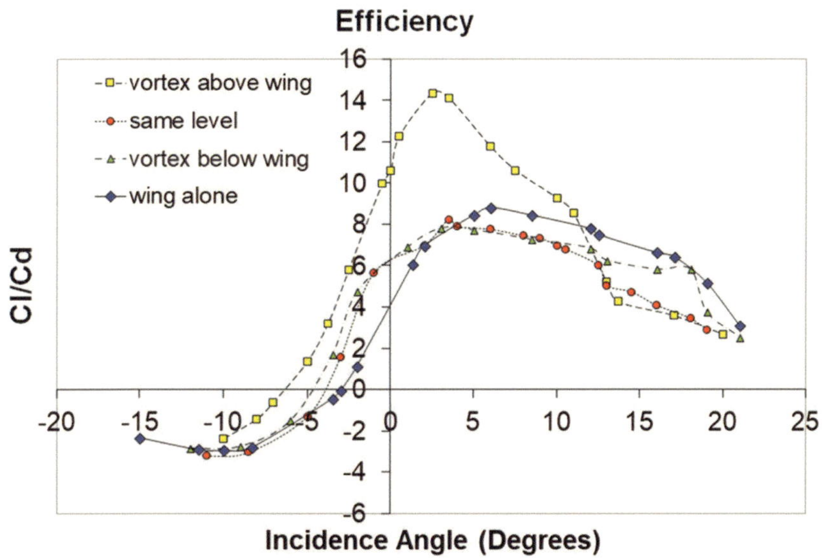

Figure 24. Efficiency of the wing in different conditions.

and not because of some extra energy present in the experiment. **Figure 24** shows clearly how a wing interacting with a vortex of its size can drastically change its performance. It is clear that the wing efficiency augments when the vortex is in the upper surface, but when the vortex is below the wing, not only the efficiency decreases, but also the stall occurs early. These kinds of result are expected since it is highly probable that the wing circulation become modified by the vorticity produced by the VG.

5. Conclusions

The main conclusions to be obtained from this work are the following:

1. As shown in Section 1, continuous wavelet transform can be used to detect a vortex passage in an anemometric measurement, even when the turbulence present in the flow has more energy than the vortex itself.
2. The characterization of any vortex is closely related to the vortex form, but some common characteristic can be obtain in any coherent structure such as radius (spatial scale) or velocity field, this can be done by detecting a passage in a time signal or by measuring the medium velocity field for longitudinal vortexes.
3. A vortex generator can modify the flow in different ways depending on the flow characteristics, as shown in the application part, a fin VG can generate a coherent structure, but for a different Reynolds number and in presence of a different geometry a fin-type VG may not generate a coherent structure, but the turbulent characteristics of the flow are always modified downstream the VG as is shown in the characterization section, and in this cases the definition of the macro and micro scales are useful.
4. It can be seen in this last experiment why the study of the vorticose structure in the free flow must be carried out, since it plays a main role in the behavior of a system, and can severely change the efficiency of such a system. Managing to characterize, generate and wisely use vortex can be helpful to modify the efficiency of a system to our will, and can be a highly beneficial approach.

Acknowledgements

We thank to bachelors Pablo Mantelli, Matias Jauregui Lorda and Gaston Santoiani for their collaboration with the measurements performed for this chapter.

Author details

Guillermo Martin Capittini[1,2*], Julio Marañon Di Leo[1,2] and Juan Sebastian Delnero[1,2]

*Address all correspondence to: guillermo.capittini@ing.unlp.edu.ar

1 UIDET-LaCLyFA, Departamento Aeronáutica, Facultad de Ingeniería, Universidad Nacional de La Plata, La Plata, Buenos Aires, Argentina

2 CONICET, Ciudad de Buenos Aires, Argentina

References

[1] Williamson CHK, Roshko A. Vortex formation in the wake of an oscillating cylinder. Journal of Fluids and Structures. 1988;**2**(4):355-381. ISSN: 0889-9746

[2] Özgören M, Sahin B, Rockwell D. Vortex structure on a delta wing at high angle of attack. AIAA Journal. 2002;**40**(2):285-292

[3] Furman A, Breitsamter C. Turbulent and unsteady flow characteristics of delta wing vortex systems. Aerospace Science and Technology. 2013;**24**(1)

[4] Study on forms of vortex breakdown over delta wing. Chinese Journal of Aeronautics. 2004;**17**(1):17

[5] Graftieaux L, Michard M, Grosjean N. Combining PIV, POD and vortex identification algorithms for the study of unsteady turbulent swirling flows. Measurement Science and Technology. 2001;**12**:1422

[6] Alekseenko SV, Kuibin PA, Okulov VL. Theory of Concentrated Vortices, an Introduction. Springer; 2007. p. 134

[7] Kadambe S, Boudreaux-Bartels GF. Application of the wavelet transform for pitch detection. IEEE Transactions on Information Theory. 1992;**38**:917-934. Print ISNN: 00189448

[8] Mallat S, Hwang WL. Singularity detection and processing with wavelets. IEEE Transactions on information Theory. 1992;**38**:617-643. Print ISNN: 00189448

[9] Abbate A, Koay J, Frankel J, Schroeder SC, Das P. Application of wavelet transform signal processor to ultrasound. Ultrasonics Symposium IEEE 1994. 1994. Print ISBN: 0-78032012-3

[10] Matlab. [Internet]. Available from: https://www.mathworks.com/help/wavelet/ref/scal2frq.html

[11] Tropea C, Yarin C, Foss JH. Handbook of Experimental Fluid Dynamics. Springer; 2007. ISBN: 978-3-540-25141-5. Chapter 10

[12] Barlow JB, Rae WH, Pope A. Low Speed wind Tunnel Testing. 3rd ed. Wiley; 1999. ISBN: 0471-55774-9. Chapter 8

[13] Boldes U, Delnero JS, Colman J, Martinez MA, Marañón Di Leo J. Lift and drag behaviour of a low Reynolds wing section submitted to a nearby stream wise vortex. AIAA. 2008;**38**: 195-200

Chapter 5

Swirl-Inducing Ducts

Trevor Frank Jones

Additional information is available at the end of the chapter

http://dx.doi.org/10.5772/intechopen.78959

Abstract

This chapter examines the flow of swirling liquid in a duct. In many cases, circumferential velocity in the cross-section of a cylindrical duct is a remarkably linear function of radius up to the proximity of the duct wall. This is similar to the behaviour of a twisting solid shaft and the analogy leads to a *solid body model* for swirl flow in ducts. Helically profiled lobate duct walls provide a twisting torque, while wall friction in simple circular ducts causes swirl to decay. The liquid counterpart of the solid body is represented as a first-order system in downstream distance because of the way torque is transmitted by duct walls rather than by shaft stiffness as in the solid case. The effect of the inertia of the rotating and twisting cylinder is unchanged from its solid counterpart, and damping is related to the viscosity of the liquid acting over the annulus between the rotating liquid cylinder and the duct wall. The shear stress in the liquid is shown to be linearly related to the intensity of the swirl. The generation of swirl is briefly described with reference to lobate designs, their development of shape and helix.

Keywords: swirl, solid-liquid pipeflow, slurry transport, computational fluid dynamics

1. Introduction

Why impart swirling flow to a stream of fluid? Spanner, a much respected naval architect, invented a helical lobate tube which increased the efficiency of heating of water in the boilers of ships [1, 2]. Importantly, his design could be manufactured economically by drawing cylindrical tube through special dies (see **Figure 1**).

For particle-bearing liquids, swirl puts particles into suspension at lower axial velocities than would be the case for a cylindrical duct. Once in suspension, particles (or debris for downstream collection) remain in full or partial suspension long after the swirl has decayed to negligible proportions.

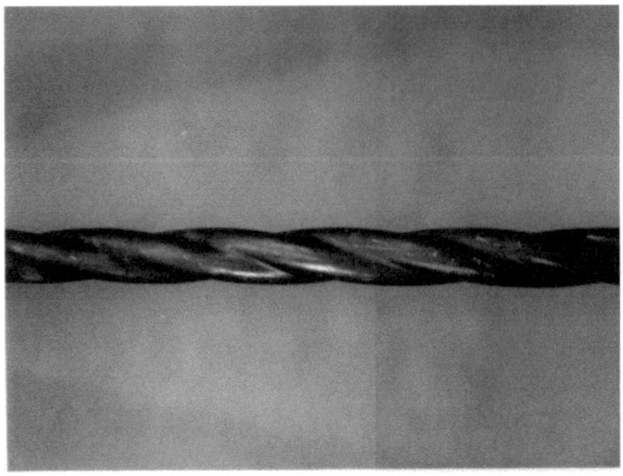

Figure 1. Three-lobe boiler tube after Spanner, 1939, 1945. Reproduced courtesy of Transport and Sedimentation Conference, Wroclaw University of Environmental and Life Sciences, Poland.

Lower axial velocity implies much lower pumping power: a strong economic reason to swirl the flow in the first place. The efficacy of swirl generation in pumping particulate liquids, particularly river slurries, has been recognized for many years. The Gordon patent for a duct with internal vanes to encourage swirl was published in 1899 [3].

Economic benefits are not the only reasons for studying swirl. In some instances, enhanced swirl is required irrespective of increased pressure losses and pumping power requirement.

The data in this chapter originate from several sources. Much of it comes from validated computational fluid dynamics (CFD) code using RANS (Reynolds-averaged Navier Stokes equations). The basis of these equations is 'Reynolds decomposition', whereby an instantaneous quantity is decomposed into time-averaged quantities and fluctuating quantities. In cylindrical polar co-ordinates (r,θ,z), the time-averaged velocities are u, v and w and the fluctuating velocities are u', v' and w'. In essence, the fluctuating quantities can be assumed to have a temporal mean of zero. This makes for an increase in the number of unknowns because the cross products ($\overline{u'u'}, \overline{v'v'}, \overline{w'w'}, \overline{u'v'}, \overline{u'w'}$, and $\overline{v'w'}$) have to be determined.

RANS turbulence modelling techniques are often classed by the number of equations used to model the flow field. Early results were obtained using two-equation $(k-\epsilon)$ and $(k-\omega)$ models, where k represents the turbulent kinetic energy, ϵ represents the rate of dissipation of turbulent energy and ω represents the specific rate of dissipation of turbulent kinetic energy into internal thermal energy. As greater computer power became available, the six-equation Reynolds stress model (RSM) became the method of choice for swirling flows. There are variations in RSM solutions. The CFX RSM-ω [5] has been used since it can give more accuracy near the wall.

2. Characterizing swirl

Firstly, I should explain what I mean by 'swirl' and then define precise ways to assess it. In turbulent pipe flow, there are many eddies and circulations cascading from the large to the

http://dx.doi.org/10.5772/intechopen.78959

small. Kolmorogov [6] showed that most of the kinetic energy in the flow is contained in large-scale structures. Energy 'cascades' to smaller scales by an inviscid mechanism until it is small enough for viscous dissipation to take place. I define swirl as large-scale, one-way circulation surrounding the geometric centre of the duct.

With a definition of swirl in place, I now need to explain the mathematical measures of this behaviour. Is it useful, strong or weak, efficient or profligate in its expenditure of pipeline pressure? The first and most obvious measure is the circumferential velocity, sometimes referred to as tangential velocity, w_{max}, taken at its maximum in the cross-section. Useful in itself, this direct measure is important in application to downstream devices—cyclones or pumps for example.

Circumferential velocity as a measure of swirl takes no account of the axial velocity required to generate or maintain it. In contrast, the *swirl angle*, incorporating the axial velocity, indicates the angular deflection of the flow and can be clearly seen in transparent pipe sections (see **Figure 2**). Tonkin [7] used these images to infer tangential velocity for a series of particle concentrations and axial velocities.

The *swirl angle*, θ_S, is given by

$$\theta_S = \tan^{-1}\frac{w_{max}}{u_m} \tag{1}$$

where u_m is the mean axial pipe velocity.

Measurement transducers can be corrupted by swirling flow, and International Standard ISO 5167 specifies a maximum swirl-angle limit of 2° at or near transducer stations.

The swirl angle does not take account of the angular momentum given to the flowing liquid. The ratio of angular momentum flux to the product of pipe radius and axial momentum flux is known as the *swirl intensity* or *swirl number*, Ω. (Note that a simple ratio of angular to axial

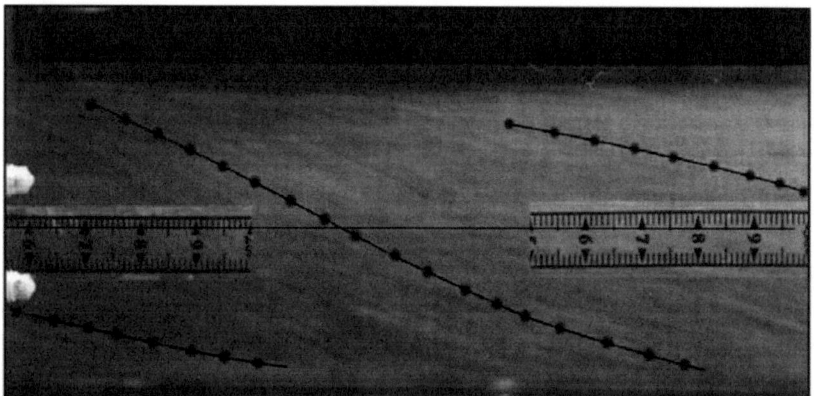

Figure 2. 1.4% by volume coarse sand in water, axial velocity 1.7 m/s showing *swirl angle*. Image from Tonkin [7] with thanks.

momentum would not be dimensionless.) There are several versions of this definition and the version defined in Eq. (2) allows for variations in the axial velocity with radial displacement.

$$\Omega = \frac{2\pi\rho \int_0^R uwr^2 dr}{R \times 2\pi\rho \int_0^R u^2 r dr} = \frac{\int_0^R uwr^2 dr}{R \int_0^R u^2 r dr} \tag{2}$$

where R is the pipe bore radius (of the cylindrical delivery pipe), u is the axial velocity at radius r, and w is the circumferential velocity at radius r.

Swirl number gives a simple way to classify swirl for computational calculation methods. If $\Omega < 0.5$, two-equation methods (such as $(k - \epsilon)$ and $(k - \omega)$) are generally considered adequate. If $\Omega \geq 0.5$, the six-equation RSM is preferred despite the increased computational cost.

Swirl intensity and swirl angle are closely related measures and in many cases an almost linear relation exists between them.

Pressure loss is an inevitable consequence of swirl generation and it is important to use that pressure effectively. Ganeshalingam [8] developed a dimensionless group, *swirl effectiveness*, S, given by

$$S = \frac{\Omega}{\frac{\Delta P}{\frac{1}{2}\rho u^2}} \tag{3}$$

where ΔP is the pressure loss over a length of duct and ρ is the fluid density.

This measure has proved invaluable in optimizing Spanner-type duct designs.

Another pressure-related metric for use when a Spanner-type duct generates swirl is the pressure loss for an equivalent length of smooth circular tube. The well-known Darcy-Weisbach equation can be used to calculate this:

$$\Delta P_s = f \frac{L}{D} \frac{\rho u^2}{2} \tag{4}$$

where f is the friction factor for a smooth duct, L is the duct length and D is the duct diameter. A simple ratio can be used to gauge the magnitude of the pressure loss suffered as a result of the use of the duct:

$$\textit{Pressure penalty factor} = \frac{\Delta P}{\Delta P_s} \tag{5}$$

3. The solid body model

In many cases of developed swirling flow, the swirl angle, θ, is found to be almost constant over most of the cross-section. **Figure 3** shows a tangential velocity profile for a nominal axial velocity

http://dx.doi.org/10.5772/intechopen.78959

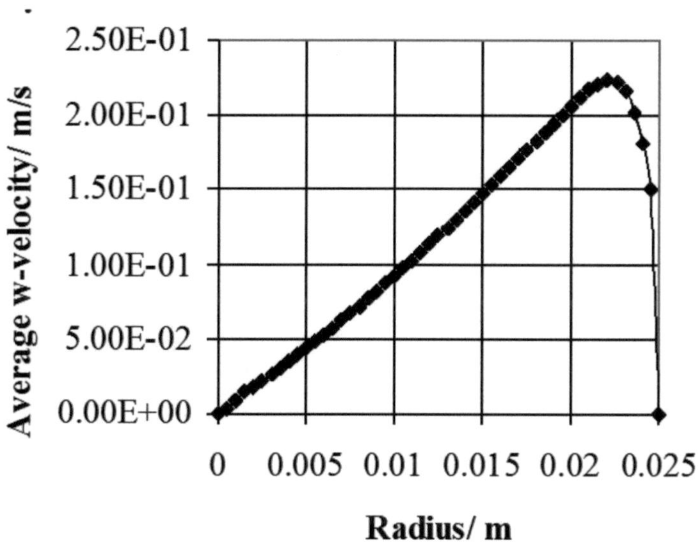

Figure 3. Circumferential velocity, w, downstream of a three-lobe swirl-inducing duct similar to Spanner's design. Data from Raylor with thanks.

of 2 m/s. The near linearity over approximately 84% of the bore also indicates that angular velocity, ω, is similarly constant in this range. Constant angular velocity is a characteristic of a solid rotating shaft and this concept suggests a simple mechanical analogy of *the solid body model* which can be used to describe swirling flow.

Damping friction (directly proportional to tangential velocity) clearly has little effect in the central 84% of **Figure 3**. The peripheral 16% of the velocity profile indicates gathering damping friction as the radius increases. At the outer radial extremity, the circumferential velocity falls to zero in accordance with the *no-slip* principle of Newtonian mechanics. This outer damping annulus is characterized by a dimensionless distance from the wall, with y^+ defined as

$$y^+ = \frac{u_* y}{\nu} \tag{6}$$

where u_* is the friction (shear) velocity $= \sqrt{\frac{\tau_W}{\rho}}$, τ_w is the wall shear stress, y is the distance to the wall and ν is the kinematic viscosity.

In turbulent pipe flow, close to the wall, is a *laminar sub-layer* of width $y^+ \sim 5$. At greater distances, up to about $y^+ \sim 35$, a *buffer layer* gradually develops the laminar sub-layer into fully turbulent flow. This is much smaller than the outer 16% of **Figure 3** and later results will show that the solid body starts at approximate distance $y^+ \sim 72$ for an axial velocity of 2 m/s. The distance is strongly influenced by the axial velocity. For axial velocities between 1 and 4 m/s in an industrial steel pipe of bore 50 mm, the distance to the wall can be expected to vary from about 25% to about 10% of the duct radius.

The simplified system dynamics of the analogy of a solid-liquid cylinder are described by three elements: the *inertia* of the shaft and any rotating components such as a flywheel, the *stiffness* of the shaft transmitting torque down the shaft and the *damping* of the speed of rotation. Straight away the stiffness of the shaft can be eliminated from the model. By definition, liquids do not have stiffness and they adapt to the shape of the containment without coercion. We are left with a shaft, length one pitch (for one 360° rotation), subjected to a torque M rotating at temporal rate $d\theta/dt$.

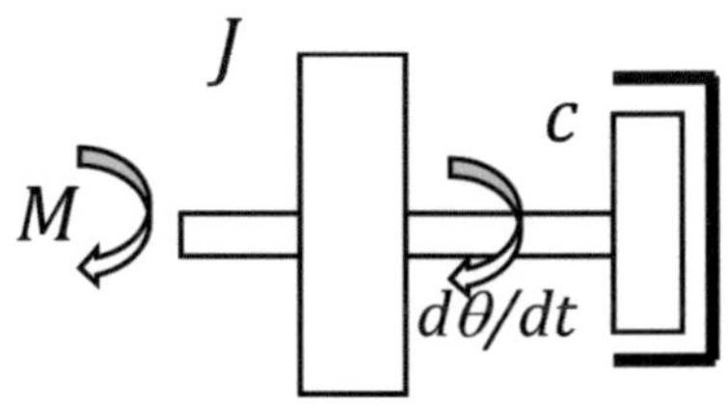

$$M = J\frac{d^2\theta}{dt^2} + c\frac{d\theta}{dt} \tag{7}$$

where J is the polar second moment of mass of the cylinder and c represents a damping coefficient dependent upon the area of the shearing surfaces. The torque moment M can be positive for a Spanner-type pipe or can approach zero for a frictionless cylindrical tube.

In the model, the coefficient of damping, c, is provided by the viscosity of the liquid. Consider fully developed swirling flow in the core of a cylindrical duct (**Figure 4**). In the example above (mean axial velocity = 2 m/s), there is a zone of approximate width $y^+ = 72$ in which all the

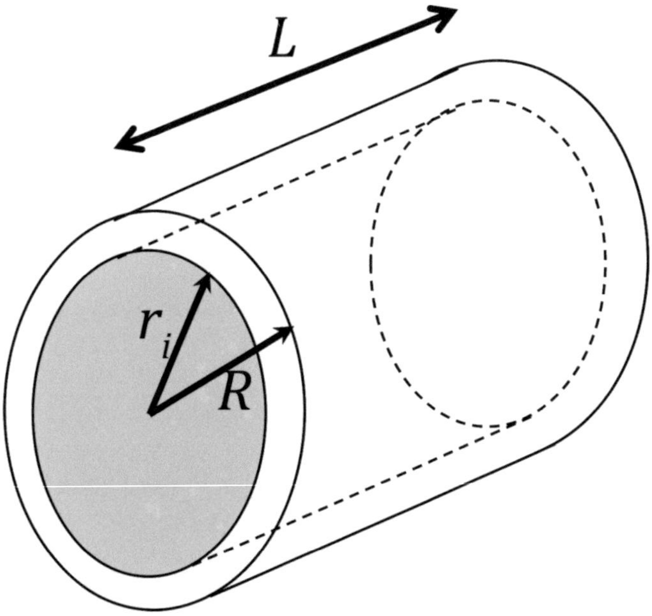

Figure 4. Fully developed swirling flow in a cylindrical duct.

http://dx.doi.org/10.5772/intechopen.78959

damping occurs. Note that if the swirl is generated by a Spanner-type duct, there will be other shearing surfaces in the core flow where the rotating annuli meet the as yet non-rotating central portion. This is sometimes referred to as a *wall-jet* profile (Steenbergen and Voskamp [9]).

In order to quantify the damping coefficient, c, I concentrate on the viscous part of Eq. (7). Put another way, I assert $\frac{d^2\theta}{dt^2} \rightarrow 0$ for the time being.

$$M_{viscous} = c\frac{d\theta}{dt} \tag{8}$$

Newton's law of viscosity gives

$$\tau = \mu\frac{du}{dy} \approx \mu\frac{r_i\frac{d\theta}{dt}}{(R - r_i)} \tag{9}$$

Torque is applied at the outer radius as wall friction or reaction from the pipe profile, so

$$\frac{M_{viscous}}{2\pi RL} = \mu\frac{r_i\frac{d\theta}{dt}}{(R - r_i)} \tag{10}$$

Comparing (10) with (8), we obtain the coefficient of damping per unit length.

$$\frac{c}{L} = \mu\frac{2\pi Rr_i}{(R - r_i)} = \mu\frac{2\pi R(R - y)}{y} \tag{11}$$

where y is the distance from the wall of the duct.

The next major challenge to the solid body model is the transmission of torque. In a solid shaft, the torque is transmitted by its stiffness, but stiffness has been discounted as a factor in liquids. In the case of a profiled swirl tube, the torque comes from the interaction of the axial flow with the walls of the tube, an interaction the author describes as the *driving function*. This becomes clear when Eq. (7) is rewritten so that the dependent variable becomes axial distance along the cylinder (z). Putting G = twist gradient $d\theta/dz$

$$M = J\frac{d^2\theta}{dz^2}\left(\frac{dz}{dt}\right)^2 + c\frac{d\theta}{dz}\left(\frac{dz}{dt}\right) = J\frac{dG}{dz}u^2 + cGu \tag{12}$$

Dividing throughout by cu

$$\frac{M}{cu} = G_D(z) = \frac{Ju}{c}\frac{dG}{dz} + G \tag{13}$$

Note that the group of variables at the left-hand side of Eq. (3) $G_D(z) = \frac{M}{cu}$ has the same dimensions as G and is the *driving function*, i.e.

$$Tu\frac{dG}{dz} + G = G_D(z) \tag{14}$$

where the time constant

$$T = \frac{J}{c} \tag{15}$$

The *complementary function*, or transient $G_{cf}(z)$, is the solution to $Tu\frac{dG}{dz} + G = 0$. This has the form

$$G_{cf}(z) = Be^{-\frac{z}{Tu}} \tag{16}$$

The solution to Eq. (14) has another (steady-state) part, the *particular integral* (PI), $G_{pi}(z)$, which depends on the driving function $G_D(z)$. The total response is a combination of these two components

$$G(z) = G_{cf}(z) + G_{pi}(z) \tag{17}$$

There are a series of driving functions of interest and I shall start with the simplest: the decay of swirl angle downstream of swirling flow from, for example, a pump output or double elbow. The driving function $G_D(z)$ in this case is a negative step change from the initial swirl angle to zero.

When $z = 0$, $G(z) = G_o$, the initial swirl gradient. Eq. (16) gives

$$B = G_o \; i.e. \; G(z) = G_o e^{-\frac{z}{Tu}} \tag{18}$$

Halsey [10] studied the swirl in clean water following a double elbow. His work was aimed at measurement devices for which swirling flow is disruptive. ISO 5167 specifies a 2° swirl-angle limit for measurement purposes and Halsey came up with an empirical law for its decay as follows

$$\theta = \theta_o e^{-\frac{1.5fz}{D}} \tag{19}$$

where θ_o is the swirl angle at commencement, θ is the swirl angle at a downstream distance z, f is the friction factor and D is the diameter of the bore. Steenbergen and Voskamp [9] arrived at an almost identical equation in terms of *swirl intensity, Ω,* instead of swirl angle. Ganeshalingam's work [8] achieved close agreement with these models and a simulation exercise (below) confirms the relationships.

Differentiating (19) gives

$$\frac{d\theta}{dz} = G(z) = \theta_o \left(\frac{-1.5f}{D}\right) e^{-\frac{1.5fz}{D}} \tag{20}$$

when $z = 0$, $G_o = \theta_o\left(\frac{-1.5f}{D}\right)$, so for the Halsey model

$$\frac{G(z)}{G_o} = e^{-\frac{1.5fz}{D}} \tag{21}$$

Equating exponents in Eqs. (18) and (21), we have a first estimate of the time constant T for decay of swirl and the length of the swirling wake Tu

$$-\frac{z}{Tu} = -\frac{1.5fz}{D} \text{ from which } T = \frac{D}{1.5fu} \text{ and } Tu = \frac{D}{1.5f} \quad (22)$$

The solid body model gives us $T = \frac{J}{c}$. The polar moment of inertia, J, of a solid cylinder, density ρ, per unit length (L) is given by

$$\frac{J}{L} = \frac{1}{2}\rho\pi\left(\frac{D}{2}\right)^4 \quad (23)$$

From this, and the time constant T, a value of c/L can be deduced which can be used with Eq. (11) to estimate the distance, y, of the solid body from the wall of the duct.

It is not possible to specify the total extinction of swirl. For some purposes, the point of 95% reduction in swirl angle (L_{95}) after a downstream distance of $3Tu$ should be a useful approximation. If swirl is a desirable property (to keep solids in suspension for example), the *half-life* distance ($L_{50} = 0.6931 \times (Tu)$) is a more appropriate concept. **Table 1** tabulates these calculations for a series of axial velocities and indicates that the level of swirl at Reynolds number of 100,000 in an industrial steel pipe with friction factor 0.022 can be assumed to have decayed to half its initial value after about 21 diameters using the solid body model.

It is generally accepted that a y^+ value of about 35 indicates the edge of the buffer layer next to the wall. The range of values of y^+ significantly greater than this value suggests that there is an annulus of turbulent flow between the buffer layer and the solid body.

The half-life distances L_{50} are approximately constant for axial pipe velocities in the range [1.0, 4 m/s]. This is an important observation for designers of pipe systems in which the axial velocity might vary.

u	Re	f	T	L_{50}	c/L	y	y^+
m/s	—	—	s	m	Nms/m	m	—
1	50,000	0.024	1.37	0.95	0.00045	0.0065	82.3
1.5	75,000	0.023	0.97	1.00	0.00064	0.0050	76.8
2	100,000	0.022	0.75	1.04	0.00082	0.0040	71.8
2.5	125,000	0.022	0.61	1.06	0.00101	0.0034	67.5
3	150,000	0.022	0.52	1.07	0.00119	0.0029	63.9
3.5	175,000	0.021	0.45	1.08	0.00137	0.0026	60.8
4	200,000	0.021	0.39	1.09	0.00156	0.0023	58.0

Clean water: pipe diameter, $D = 0.05$ m; roughness height, $\varepsilon = 0.000046$ m; relative roughness, $\varepsilon/D = 0.00092$; polar moment of inertia per m, $J/L = 0.000614$ kgm^2/m.

Table 1. Solid body model: industrial steel pipe transporting clean water.

3.1. Wall shear stress in the solid body model

Starting from an analysis by Kitoh [4], the tangential momentum equation for axi-symmetric flow gives an equation for circumferential shear stress at the wall in the decay of swirl in a circular pipe.

$$\tau_{r\theta} = \rho v w + \frac{\rho}{R^2}\int_0^R r^2 \frac{\partial}{\partial z}\left(uw + \overline{u'w'} - \nu\frac{\partial w}{\partial z}\right)dr \tag{24}$$

Now $\left(uw \gg \overline{u'w'} - \nu\frac{\partial w}{\partial z}\right)$ and putting $\rho v w \to 0$, Eq. (24) for the wall can be written

$$\tau_w = \frac{\rho}{R^2}\int_0^R r^2 \frac{\partial}{\partial z}(uw)dr \tag{25}$$

where τ_W is the circumferential wall shear stress.

Leibnitz's rule for the differentiation of integrals allows the change of order of integration and differentiation in Eq. (25):

$$\tau_W = \frac{\rho}{R^2}\frac{d}{dz}\int_0^R r^2 uw dr \tag{26}$$

For a constant Reynolds number, the axial velocity $u = u_m$ is constant and the axial momentum can be simplified to a constant quantity:

$$2\pi\rho\int_0^R u^2 r dr = 2\pi\rho u_m^2\int_0^R r dr = 2\pi\rho u_m^2\frac{R^2}{2} = \pi\rho R^2 u_m^2 \tag{27}$$

This allows the simplification of swirl intensity to

$$\Omega = \frac{2\pi\rho\int_0^R r^2 uw dr}{R \times \pi\rho R^2 u_m^2} = \frac{2\int_0^R r^2 uw dr}{R^3 u_m^2} \tag{28}$$

and substituting for Ω in Eq. (26)

$$\tau_W = \frac{\rho}{R^2}\frac{d}{dz}\int_0^R \left(\frac{\Omega u_m^2 R^3}{2}\right)dr = \frac{1}{2}\rho u_m^2 R\frac{d\Omega}{dz} \tag{29}$$

i.e.

$$\frac{\tau_W}{\frac{1}{2}\rho u_m^2} = R\frac{d\Omega}{dz} = \frac{1}{2}\frac{d\Omega}{d\frac{z}{D}} \tag{30}$$

Note that since $\frac{d\Omega}{d\frac{z}{D}}$ is always negative in decaying swirl flow, values of resisting shear stress τ_W must be given a negative sign.

Returning to the analogy of a solid body for the flow, one might reasonably expect a linear relationship between circumferential stress and circumferential strain (swirl intensity or swirl angle) for a given Reynolds number. This can be tested with a straightforward simulation experiment.

The simulation experiment (below) gives $\Omega = \Omega_0 e^{-0.0338\left(\frac{z}{D}\right)}$ for the object 50-mm smooth tube. Factoring in the measured mean friction factor yields $\Omega = \Omega_0 e^{-\xi f\left(\frac{z}{D}\right)}$ where $\xi = 1.711$.

The imposition of pipe roughness considerably increases the friction factor, f. For an axial velocity of 1.64 m/s in a commercial steel pipe (ε = 0.000046 m), friction factor is sharply increased to 0.0228. If the stress-strain assumption still holds, substituting this increased friction factor in Eq. (31) yields $-1.483f \times \Omega = \frac{d\Omega}{d\frac{z}{D}}$. The constant $\xi = 1.483$ is very close to that proposed by Steenbergen and Voskamp who achieved $\xi = 1.49 \pm 0.07$ for a larger range of values $0 \le \Omega \le 0.18$ [9].

SIMULATION EXPERIMENT: swirl decay in a cylindrical tube

Figure 5 shows the results of a simple RANS simulation for the flow of clean water through a 50-mm diameter smooth circular tube using the Reynolds stress model ($-\omega$ version) [5]. The entry plane is furnished with a mean axial velocity of 1.64 m/s and an initial circumferential velocity of 0.72 m/s at the wall and zero at the centre of the tube. The boundary conditions including the law of the wall are allowed to develop over an axial length of 10.0 m. The solid body model implies that wall friction will have reached 95% of its effect on the flow after $3Tu$ m and 99% after $5Tu$ m.

Reynolds number $Re = \frac{\rho u D}{\mu} = 997 \times 1.64 \times \frac{0.05}{0.001} = 81,754$

Friction factor (Blasius equation): $f = 0.3162 \times Re^{-0.25} = 0.0186$

Time constant (Eq. (22)) $T = \frac{D}{1.5fu} = \frac{0.05}{1.5 \times 0.0186 \times 1.64} = 1.09$ s

Effective range downstream: $9.5\ \text{m} \le z \le 9.98\ \text{m}$

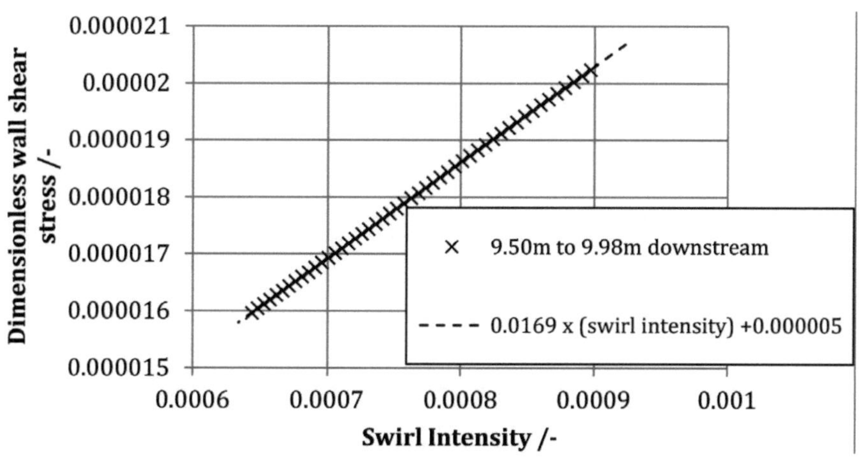

Figure 5. Dimensionless wall shear stress plotted against swirl intensity.

For the solid body model this is $5.30Tu \leq z \leq 5.57Tu$

Measured mean friction factor over range: 0.0197

In this range, the regression law applied to the CFD data is very precise ($r^2 = 0.99999$). Ignoring the small intercept as an enabling assumption,

$$-0.0169\Omega = \frac{1}{2}\frac{d\Omega}{d\frac{z}{D}} \tag{31}$$

$$\Omega = \Omega_0 e^{-k\left(\frac{z}{D}\right)} \tag{32}$$

where $k = 0.0338$ is a constant of proportionality and Ω_0 is the swirl intensity at outset.

Note the significant difference in time constant for smooth pipe when compared to commercial industrial steel pipe because of the increased value of the multiplier ξ. In the example, the time constant is reduced from 1.09 (smooth pipe) to 0.905 s (industrial steel pipe).

Since the swirl angle is linearly related to swirl intensity in most cases, it follows that Halsey's correlation [10] also fits the data.

4. Generating swirl

Previously, we have seen that a solid body model can be applied to the simple case of swirl decaying downstream. In these cases, the driving function is simply a step to zero: $G_D(z) \rightarrow 0$. We now come to more complex situations where the goal is to generate swirl for a specific purpose. Before addressing this problem, we must first consider the cross-sectional shape of the duct to be twisted.

The contours of duct walls should impart torque to the flow while minimizing pressure loss. By designing using this criterion, pressure costs are used in an effective way. Here, Ganeshalingham's dimensionless group, *swirl effectiveness*, *S*, can be used to evaluate the effectiveness of swirl generation.

$$S = \frac{S}{\frac{1}{2}\rho u^2} \tag{33}$$

4.1. Lobate designs

The boiler tube patented by Spanner and illustrated in **Figure 1** has only three lobes. Raylor [11] idealized the lobe profiles to form semicircular shapes for his CFD modelling to test the design for the transportation of particle-bearing liquids. The computer modelling was underpinned by experimental work on an extant boiler tube. Later work by Ganeshalingam showed that a four-lobe duct (or a 2-lobe duct) was more efficient when compared on the basis of swirl effectiveness.

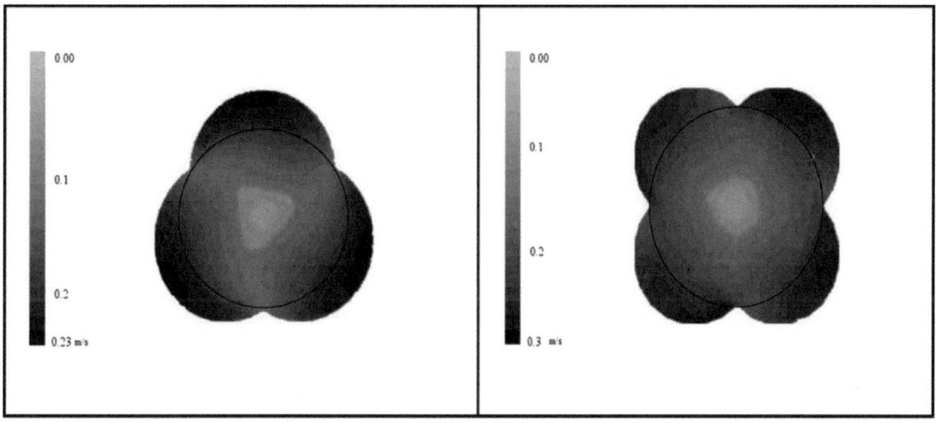

Figure 6. Contours of tangential velocity in three-lobe and four-lobe swirl pipes. Pure water with an axial velocity 2 m/s. Source: Jones and Ariyaratne [12]; reproduced courtesy: AIChE Journal.

In **Figure 6**, after Ariyaratne [13], it can be seen that the contours of tangential velocity adopt a more circular pattern in the four-lobe variant and that an efficient circulating core flow is produced in consequence.

Simply equating the area of the four-lobe duct to πR^2, the upstream area, the ratio of the lobe radius, R_4, to the upstream radius can be derived:

$$\frac{R_4}{R} = \sqrt{\frac{\pi}{2\pi + 4}} = 0.5527 \tag{34}$$

R_4 and sub-multiples have been used in other speculative swirl-duct designs so that lobe sizes can be compared across designs.

4.2. Response of the solid body model to a constant-pitch swirl duct

I first consider a four-lobe swirl duct with constant pitch:diameter ratio of 8:1 simply connected in line after a cylindrical duct. The driving function for this is a positive step or Heaviside function in swirl gradient $\frac{d\theta}{dz}$. If the imposed value of swirl gradient is G_0, this can be tried as the particular integral in Eq. (17):

$$G(z) = G_{pi}(z) + G_{cf}(z) = G_0 + Be^{-\frac{z}{Tu}} \tag{35}$$

The constant B can be easily obtained by considering the boundary condition $G(0)$ which yields $B = -G_0$. So, the solution for this case is satisfied by

$$G(z) = G_0\left(1 - e^{-\frac{z}{Tu}}\right) \tag{36}$$

Figure 7 illustrates the response of a system comprising a four-lobe Spanner-type duct with cross-sectional area equal to a cylindrical upstream main of diameter 50 mm carrying clean

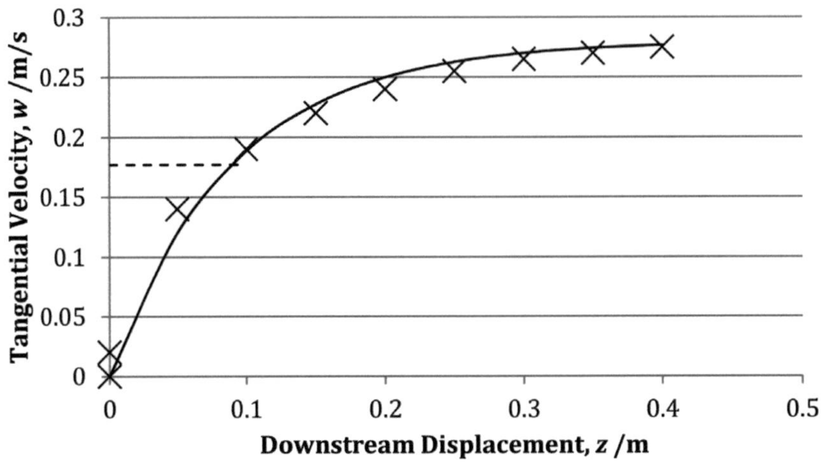

Figure 7. Response of tangential velocity in a Spanner-type swirl-inducing duct of diameter 50 mm carrying clean water at 2 m/s. Radial position is 0.7*R*. Reproduced courtesy of Transport and Sedimentation Conference, Wroclaw University of Environmental and Life Sciences, Poland.

water at an axial velocity of 2 m/s. The ordinates are tangential velocities at a radius of 0.7*R*. From these ordinates, the response of *G*(*z*) can be deduced given that the twist gradient of the duct is 15.71 radians/m. Fitting Eq. (36) to these data yields $Tu = 0.1$ m. In the first-order system, 63.2% of the final value is reached after *Tu* metres.

The length of the wake (the point at which swirl has decayed by 95%) is 3*Tu*. Hence the standard length of this four-lobe Spanner-type duct with axial velocity 2 m/s is 300 mm.

Note the apparent anomaly between the calculated constant *Tu* for the swirl generation case (0.1 m) and that for the swirl-decaying case from **Table 1** (1.5 m), a multiplicative factor of 15. The time constant of the model must be identical whatever driving function is applied so the only way in which this phenomenon can be explained is by studying the changes in the geometry of the rotating core for the generation case. The circumferential velocity distribution for swirl generation is a *wall jet* [9] in which circumferential velocity is concentrated towards the outer perimeter of the core flow. Effectively, the solid body in this case is not a solid cylinder, but more like a cylindrical tube. Thinking of the time constant of the system ($T = \frac{J}{c}$), the polar moment of inertia, *J*, is significantly reduced from that of a solid cylinder while the shearing surfaces are significantly greater than those of a solid cylinder, increasing the damping coefficient *c*.

4.3. Cross-section development for lobate ducts

The example of a fixed-pitch duct is useful in that it gives a standard length for swirl pipe designs. However, a lobate swirl duct cannot be added directly to a cylindrical pipe without incurring wasteful pressure losses. A better solution is to allow the shape to develop in a sigmoidal fashion. A family of sigmoidal coefficients is given by

http://dx.doi.org/10.5772/intechopen.78959

$$\beta\left(\frac{z}{L}\right) = \left[\frac{1 - \cos\left(\pi \frac{z}{L}\right)}{2}\right]^n \tag{37}$$

and illustrated in **Figure 8**.

For a Spanner-type lobate duct, the sigmoidal function can be used to schedule the growth of lobe area, the expansion of the duct or, usually, the development of the radius of the lobe to its final value. In this case, the factors are

$$\beta\left(\frac{L-z}{L}\right) \textit{ for entry, } \left\{1 - \beta\left(\frac{L-z}{L}\right)\right\} \textit{ for exit} \tag{38}$$

The exponent n governs the steepness of the sigmoidal curve. If $n > 0.5$, the initial gradient is effectively zero, giving a gradual increase in shape.

Originally, a *Three-Zone* development was proposed by Jones and Ariyaratne [12] whereby the entry section was a beta transition with $n = 2$, followed by a fixed-pitch section, followed again by an exit transition with $n = 0.5$. This was certainly an improvement on the system without transitions, but the fixed-pitch section constrained the angular acceleration of the liquid in the duct. Later designs embodied a continuous development of shape and helix. The *symmetric* development had an increasing beta function to the central point of the duct followed by a decreasing beta function to the outlet. The *asymmetric* development had an increasing beta function to a point two-thirds along the length of the duct and a decreasing beta function for the latter third. The asymmetric development gave slightly better swirl intensity results than the symmetric development.

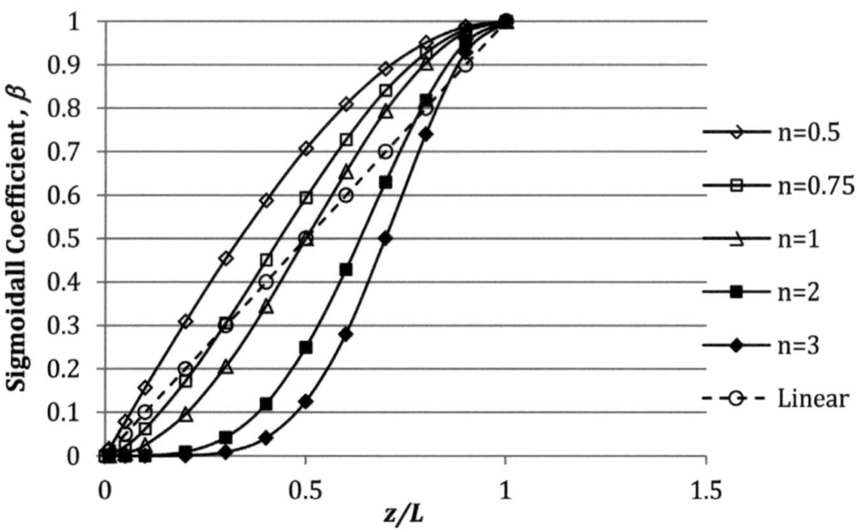

Figure 8. Coefficients for the development of cross-section shape.

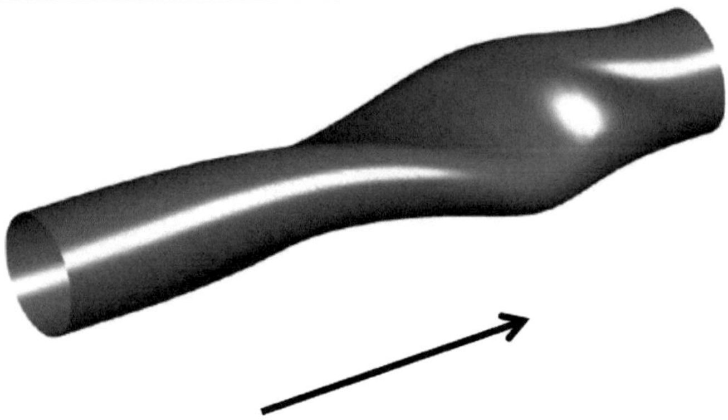

Design variables		Performance	
Upstream diameter	50 mm	Axial velocity	1.5 m/s
Length	300 mm	Reynolds number	74,775
Cross-section	2-lobe	Friction factor[a]	0.0191
Lobe radius	13.8 mm	$\Delta P^{**} =$	333.25 Pa
Shape development	Asymmetric 2:1	Cylindrical tube[c] $\Delta P_s =$	128.6 Pa
Entry sigmoid	$n = 2$	Pressure penalty	2.59
Exit sigmoid	$n = 2$	Swirl intensity at outlet plane[b] (swirl number)	0.0716
Total twist	180°	Swirl effectiveness[b]	0.241
Helix development	$\lambda z + G_0$	[a]Blasius equation for smooth tubes: $f = 0.3162 \times Re^{-0.25}$	
$G_0 =$	0.0	[b]RSM-ω [5]	
$\lambda =$	69.81 rad/m/m	[c]Darcy-Weisbach equation: $\Delta P = f \frac{L}{D} \frac{\rho u^2}{2}$	

Table 2. Design data sheet for a two-lobe swirl inducing duct (lobe radius = 1 × R_4).

4.4. Helix development

Raylor [11] showed that advantages accrued from the gradual angular acceleration of twist in a profiled tube. In recent work, this has been combined with the asymmetric beta function to create a duct with developing cross-sections and acceleration of twist throughout the tube. Inserting the driving function for this case we obtain

$$Tu\frac{dG}{dz} + G = \lambda z + G_0 \tag{39}$$

where λ is the rate of increase of G along the duct, and G_0 is the starting value.

Eq. (16) gives the complementary function as in the previous cases

$$G_{cf}(z) = Be^{-\frac{z}{Tu}} \tag{40}$$

http://dx.doi.org/10.5772/intechopen.78959

The *particular integral* (PI) is obtained by trying $G(z) = Pz + Q$ in [4] and equating to the ramp function $G_D(z) = \lambda z + G_0$

$$TuP + Pz + Q = \lambda z + G_0 \tag{41}$$

Comparing coefficients we have the solution for the PI

$$G_{pi}(z) = \lambda z - Tu\lambda + G_0 \tag{42}$$

Hence, the complete solution (PI + CF) is given by

$$G(z) = G_{pi}(z) + G_{cf}(z) = \lambda z - Tu\lambda + Be^{-\frac{z}{Tu}} + G_0 \tag{43}$$

Applying the boundary condition $G(0) = G_0$ gives $B = Tu\lambda - G_0$.

So

$$G(z) = \frac{d\theta}{dz} = \lambda z - Tu\lambda + (Tu\lambda - G_0)e^{-\frac{z}{Tu}} \tag{44}$$

Eq. (44) specifies the response of the solid body model to the ramped driving function in Eq. (39). A two-lobe design is illustrated in **Table 2**. The design is a modestly twisting tube but pressure losses are considerably larger than those expected in a smooth straight duct for the same duty (using the Darcy-Weisbach equation for this prediction). Increasing the amount of twist and increasing the number of lobes to four can improve the performance of the tube at the expense of increased pressure loss.

5. Conclusion

The purpose of this chapter has been to examine the technical aspects of swirling flows and to facilitate the design of ducts for specific purposes. Swirling flow is a complex, while stunningly beautiful, phenomenon and my work has been guided by the need to reduce its complexity for the designer. The emphasis has been on Spanner-type profiled tubes, but this is by no means the only way to generate swirl. The fascinating medical prospect that small amplitude helically coiled pipes might be used as bypass grafts to prevent occlusion by thrombosis has been the subject of scholarly study [14, 15].

The efficacy of the first-order solid body model was demonstrated by the simulation of flow through a 10.0-m cylindrical tube. The prediction that the downstream data taken after a distance of $3Tu$ would be representative of fully developed flow was amply demonstrated. Swirling flows are a little more difficult to predict than by using a simple exponential decay formula, but the solid body model introduced in this chapter is a simple and useful tool to apply to the design task.

Acknowledgements

As always, I am indebted to my research students Benjamin Raylor, Jeyakumar Ganeshalingam, Chanchala Ariyaratne and Ruth Tonkin for their tireless experimental and computational work in the early days. I am particularly indebted to Benjamin Raylor for his continued efforts to the present day, his enthusiasm for swirl ducts, his hard work and unfailing support.

Author details

Trevor Frank Jones

Address all correspondence to: info@tfjconsulting.com

TFJ Consulting Ltd., Derby, England

References

[1] Spanner EF. British Patent GB521548; May 24, 1940

[2] Spanner EF. British Patent GB569000; April 30, 1945

[3] Gordon HM, HA. Conduit or pipe. US Patent 630,605; August 8, 1899

[4] Kitoh O. Experimental study of turbulent swirling flow in a straight pipe. Journal of Fluid Mechanics. 1991;**225**:445-479

[5] ANSYS. Omega Reynolds Stress model. ANSYS CFX-Solver Theory Guide, Release 17. Canonsburg, PA: ANSYS Inc. pp. 95-96

[6] Kolmorogov AN. On degeneration (decay) of isotropic turbulence in incompressible viscous fluids. Doklady Akademii Nauk SSSR. 1941;**31**:538-540

[7] Tonkin RJJ. Swirling pipeflow of non-Newtonian and particle-laden fluids [thesis]. University of Nottingham; 2004

[8] Ganeshalingam J. Swirl induction for improved solid-liquid flow in pipes [thesis]. University of Nottingham; 2002

[9] Steenbergen W, Voskamp J. The rate of decay of swirl in turbulent pipe flow. Flow Measurement and Instrumentation. 1998;**9**:67-78

[10] Halsey DM. Flowmeters in swirling flows. Journal of Physics E: Scientific Instruments. 1987;**20**

[11] Raylor B. Pipe design for improved particle distribution and improved wear [thesis]. University of Nottingham; 1998

[12] Jones TF, Ariyaratne C. Design and optimisation of swirl pipe geometry for particle-laden liquids. AICHE Journal. 2007;**53**(4):757-768

[13] Ariyaratne C. Design and optimization of swirl pipes and transition geometries for slurry transport. University of Nottingham; 2005

[14] Cookson AN. Computational investigation of helical pipe geometries from a mixing perspective [thesis]. London: Department of Aeronautics, Imperial College; 2009

[15] Caro CG, Seneviratne A, Heraty KB, Monaco C, Burke MG, Krams R, et al. Intimal hyperplasia following implantation of helical-centreline and straight centerline stents in common carotid arteries in healthy pigs: Influence of intraluminal flow. Journal of the Royal Society Interface. 2013;**10**. DOI: 10.1098/ rsif 2013.0578

Chapter 6

Particle Adhesion in Cyclone Particle Separators

Yuanye Zhou and Shan Zhong

Additional information is available at the end of the chapter

http://dx.doi.org/10.5772/intechopen.79902

Abstract

Cyclone particle separators use centrifugal forces generated by swirling flow to separate particles from particle-laden flow. They are one of the key components of vacuum cleaners. However, fine particles tend to adhere on the internal wall of the cyclone and, in the worst case; this can cause severe blockage of the cyclone exit. Interactions between particles and a solid wall in cyclone are very complex and affected by many factors. Thus, the mechanism of particle adhesion is poorly understood. This book chapter will begin with a brief introduction of the working principle and common configurations of cyclone particle separators. The characteristics of particle adhesion patterns in a cyclone particle separator will then be presented and the mechanisms will be discussed based on our experimental results. After this, an experimental study supplemented by CFD simulations aiming to understand the effect of particle inlet positions on particle adhesion will be reported. Finally, a 2D numerical simulation which models interactions of particles with a solid wall and provides an insight of the key parameters that influence the particle adhesion process will be described. The finding from this work will benefit the design of cyclone particle separators for vacuum cleaners.

Keywords: cyclone particle separator, particle adhesion

1. Introduction

Cyclone particle separators (in short for cyclone) are widely used in the particle processing technology. A typical cyclone consists of a tangential inlet, a cone shape body, and a vortex finder at the top centre, as shown in **Figure 1**. The particle-laden flow enters the cyclone through the rectangular tangential inlet, forming a swirling flow. Particles are separated to the internal wall of cyclone due to the centrifugal force and are collected in the dust collector located below the cyclone. Clean air leaves the cyclone through the vortex finder.

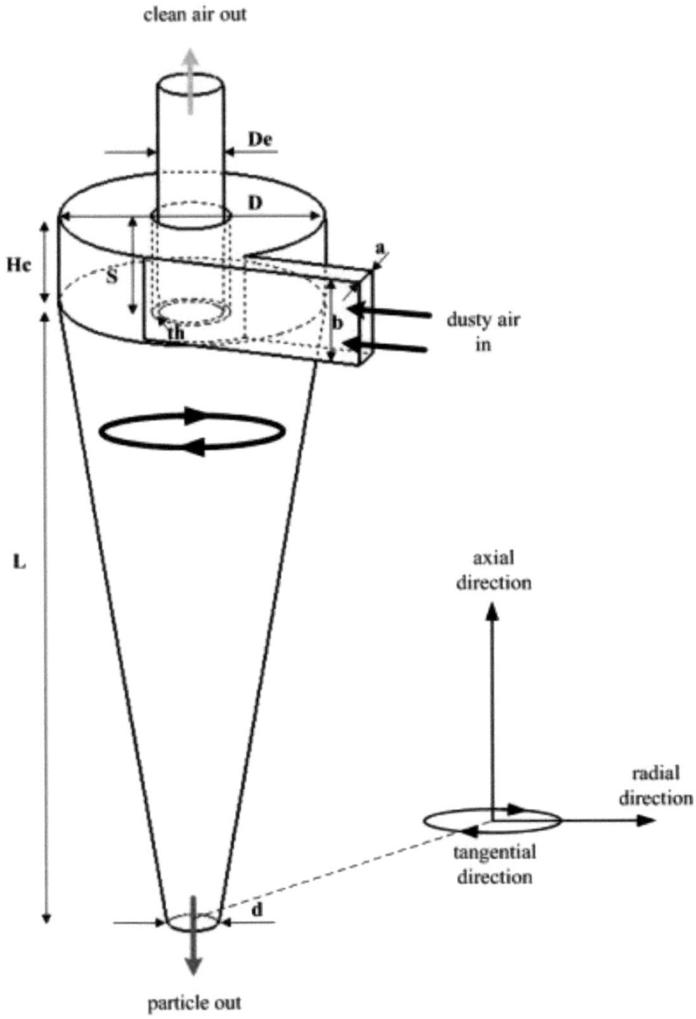

Figure 1. Illustration of a typical cyclone and its key dimensions [6].

It is important to avoid particle adhesion in the cyclone, as particle adhesion can cause the blockage of the cyclone, which deteriorates the performance of the cyclone [1]. Theoretically, particle adhesion in the cyclone is affected by the capillary force, the van der Waals force, the electrostatic force and the aerodynamic force. The parameters that affect these forces are the material properties, air flow velocity in cyclone and the environment conditions. For example, the surface energy determines the van de Waals force between particle and surface; the humidity in the air affects the capillary force; the static charge of particle influences the electrostatic force; and the air flow velocity determines the aerodynamic force [2].

As the database of particle properties only includes certain kind of particles, it is extremely difficult to find the value of surface energy and level of particle charge for particles that are not in the database. In addition, the air flow velocity distribution is complicated in the cyclone. In the cyclone, there is a fundamental flow structure, named as 'Rankine' vortex, which has a solid vortex core and a free rotational outer vortex. Besides, there are several secondary flows in cyclone, such as the secondary flow under the roof and the secondary flow along the wall in axial direction. Moreover, since the air flow in cyclone is naturally unsteady, there exists a precessing vortex core (PVC) phenomenon [3], which is a large-scale coherent structure in cyclone. At the cyclone tip region, the PVC phenomenon is the precessional bent vortex end (PBVE) attached to the wall surface [4]. It is found that the particle concentration also affected the local air flow velocity [5]. Moreover, cyclone usually deals with non-sphere particles, which requires models to predict forces on non-sphere particles. Thus, the interactions between particles and solid wall and flow conditions in cyclone are very complex. It remains unclear about mechanisms of particle adhesion in the cyclone.

This book chapter will present works, including experimental investigations of the particle adhesion patterns in the cyclone, experimental and CFD studies of particle inlet position on particle adhesion, and 2D numerical modelling of key parameters that influence the particle adhesion. The finding from this work will benefit the design of cyclone particle separators for vacuum cleaners.

2. Experimental investigation and numerical modelling of particle adhesion in cyclone

Particles and cyclones used in this book chapter were described. The particles used in the experiment were plaster particles (Thistle Dura-Finish plaster, manufactured by British Gypsum Ltd.). The average dynamically equivalent diameter of plaster particles was 1.13 μm. The density of this plaster particle was 2300 kg/m^3. The dimensions of cyclones were given in **Table 1**. The material to make the cyclones was Acrylonitrile Butadiene Styrene (ABS). The surface roughness of conical part of cyclone was 7.63–7.85 μm.

2.1. Characteristics and mechanisms of particle adhesion patterns in cyclone

In this section, characteristics and mechanisms of particle adhesion patterns were studied Particle adhesion patterns were visualised during and after the experiment and explanations for experimental observations were given.

Feature	Dimension, mm	Feature	Dimension, mm
a	5	S	13.55
b	11.6	Hc	12.1
D	35	L	87.11
De	8.54	d	6.6
th	1		

Table 1. Dimensions of cyclones used in the experiment.

2.1.1. Experiment setup

The test rig to investigate particle adhesion patterns consisted of a triboelectric charge generator (TCG), a cyclone, a fibre filter, an air pump with a pump voltage adjustor and sensors for pressure and air mass flow rate measurement. A detailed description of test rig can be found in [6]. The air mass flow rate of the cyclone was fixed at 2.1 g/s. Four particle load rates were tested. They were 0.28, 0.60, 0.75, and 0.96 g/m^3, respectively. The duration of each test was 20 min. The environmental temperature and humidity during the experiment was $20 \pm 2°C$ and $50 \pm 5\%$ RH, respectively. The variation of room pressure was less than 5% of 101 kPa.

2.1.2. Experiment results

After the experiment, pictures of particle adhesion patterns were taken. It was found that there were three types of particle adhesion patterns in cyclone. One was large-scale spiral patterns (SPs), one was small-scale wave patterns (WPs) and the last one was the thick adhesion layer (TAL), which was within 10 mm away from cyclone tip, as shown in **Figure 2**. Large-scale SPs were spiral lines that started from the upper part of the conical part and ended at the cyclone tip. It was found that SPs was caused by the spiral particle trajectory in cyclone [6]. Moreover, the TAL was a sudden increment of thickness of particle adhesion near the cyclone tip. It was believed that the cause for the TAL was due to the low wall shear stress induced by the precessional bent vortex end (PBVE) at the cyclone tip [6]. In addition, small-scale WPs were found to be embedded in the large-scale SPs. The WPs were similar to droplet patterns and chevron patterns [7, 8]. They were small-scale approximate periodic discrete patterns in the cyclone.

By using a transparent conical part of the cyclone, the development of particle adhesion patterns was visualised during the experiment at the highest particle load rate condition (0.96 g/m^3). The WPs was found to creep against the air flow direction, as shown in **Figure 3**. The white mark window tracked the same individual WP and the vertical arrow was fixed as a reference location. Comparing different frames in **Figure 3**, the relative location between the mark window and reference arrow had been changed. It meant that the individual WP moved against the air flow direction, because the air flow direction was from the right to left.

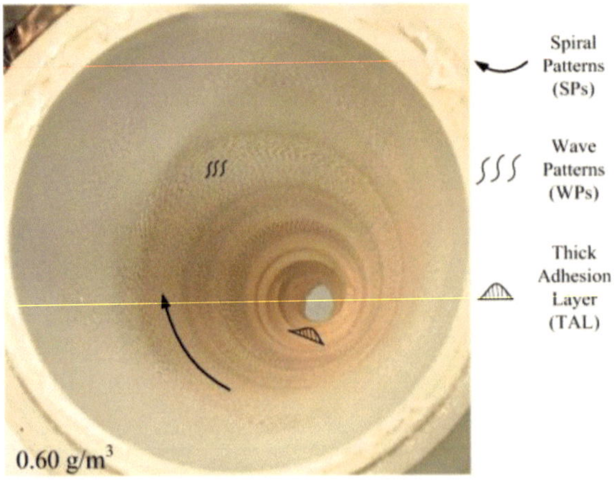

Figure 2. Key features of particle adhesion patterns in cyclone [6].

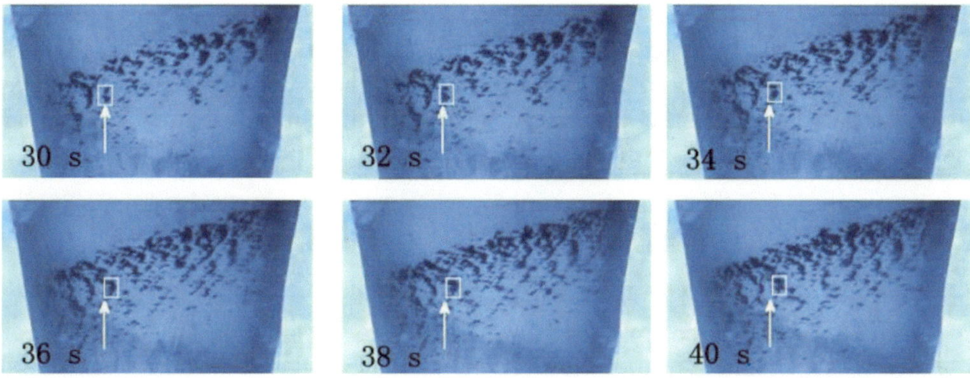

Figure 3. Backward creeping motion of the WPs in the cyclone at particle load rate 0.96 g/m^3 (air mass flow rate 2.1 g/s) [6].

However, the creeping velocity of the WP was slow, as it only moved less than 2 mm in 10 s, which meant the creeping velocity was less than 0.2 mm/s. For other WPs, similar motions were found as well.

2.1.3. Discussion

The formation of SPs and TAL was explained by the spiral particle trajectory and weak wall shear stress induced by PBVE, respectively [6]. In addition, a hypothesis was proposed to explain the formation of WPs. The mechanism of the WPs creeping against the air flow direction was comparable to the mechanism of slowly moving sand dune, as shown in **Figure 4**. The movement of the WPs and sand dune was affected by the particle deposition and removal. For the

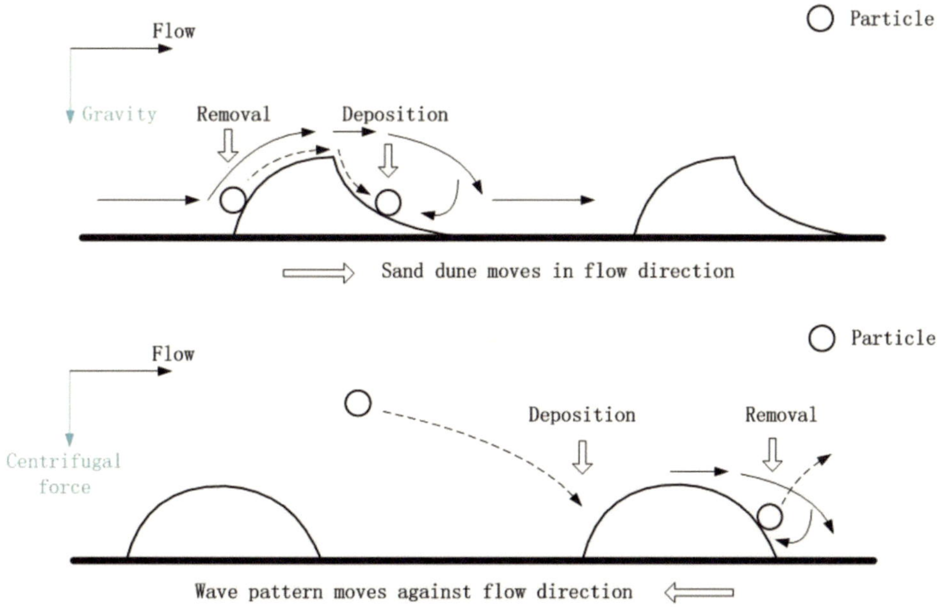

Figure 4. Sketch showing the mechanism of the backward creeping motion of the WPs [6].

sand dune, the sand particles were removed on the windward side and deposited on the leeward side. Therefore, the sand dune slowly moved in the air flow direction. However, for the cyclone, the particles deposited on the windward side and were removed on the leeward side (may be due to flow separation). Thus, the WPs moved against air flow direction. Since the accumulation of particles on the windward side of WPs required time, the WPs moved slowly.

2.2. Effect of particle inlet position on particle adhesion in cyclone

In this section, studies on the effect of particle inlet position on particle adhesion in cyclone were presented. Two different inlet parts were tested in experiment and CFD simulation. Combining the experimental results with CFD simulation results, a hypothesis was proposed to understand the effect of partly blocked inlet on particle adhesion.

2.2.1. Experiment setup

As the cyclone was small, which made it difficult to directly inject particles from a particular position, we studied the effect of partly blocked inlet on particle adhesion in the same test rig as described in previous section, except for the inlet part. Two different inlet parts were tested. They were made from an original inlet by blocking the upper left channel and lower left channel, respectively, as shown in **Figure 5**. The air mass flow rate of the cyclone was fixed at 2.1 g/s. Four particle load rates were tested 'low', 'medium', 'high', and 'very high'. They were 0.21, 0.49, 0.68, and 0.89 g/m^3, respectively. Particle adhesion and pressure drop of cyclone were measured.

2.2.2. CFD simulation setup

The CFD simulation was conducted in the software of Star-CCM+ by using unsteady Reynolds Stress Model (URSM). Structured meshes were generated and the total number of mesh cells was around 3 million, which satisfied the mesh sensitivity study. Boundary conditions of the CFD simulation were given in **Table 2**. Based on these boundary conditions, the simulated air mass flow rate was 2.15 $\pm$ 0.1 g/s that was close to the experimental condition (2.1 g/s). The density of particle in CFD simulation was the same as experiment. The geometry of particle was sphere with a constant diameter of 1.13 μm, which was the same as average particle size in experiment. After obtaining a stable flow field of the cyclone, particles were evenly distributed and injected at the inlet with an injection rate of 108 per time step. A detailed description of equations for particle simulation can be found in [6].

2.2.3. Experimental results

After experiment, the measured weight of the particle adhesion showed that the particle adhesion on the cyclone with the 'I-block' inlet was always less than that of the 'II-block' inlet (except for the 'low level', when the weight was close to zero), as shown in **Figure 6**. The difference of the weight of the particle adhesion between the 'I-block' inlet and 'II-block' inlet was around 80% at different particle load rate levels.

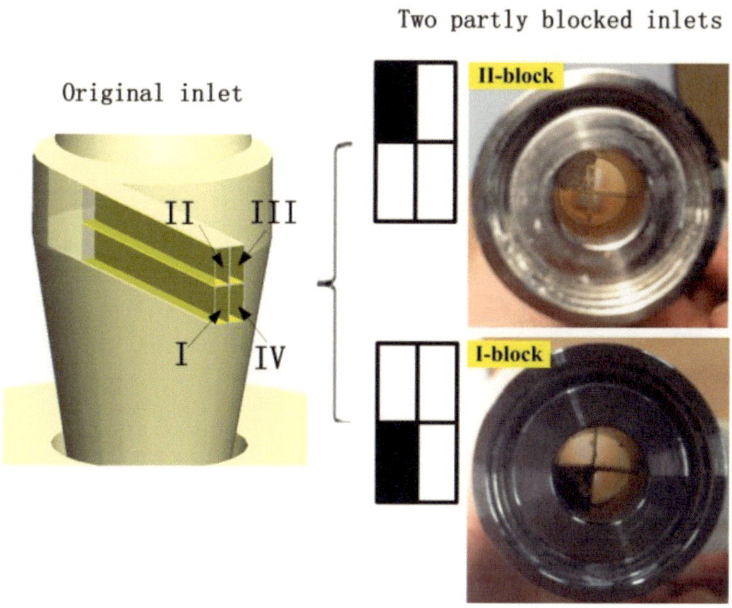

Figure 5. Two partly blocked inlets made from an original inlet.

Boundary conditions	
Inlet total pressure, kPa	0
Outlet static pressure, kPa	−9
Time step, μs	21
Atmospheric pressure, kPa	101

Table 2. Boundary conditions of CFD simulation.

Results of the time average pressure drop of cyclone over 20 min are shown in **Figure 7**. Under different particle loading rates, the time average pressure drop of the 'I-block' inlet was always higher than that of the 'II-block' inlet. The difference was around 1500 Pa. Thus, the flows with the particles loading were different between the 'I-block' and 'II-block' inlets.

It was clearly that the 'I-block' inlet always had less particle adhesion and higher pressure drop, compared with the 'II-block' inlet.

2.2.4. CFD simulation results

The particle trajectory was simulated in the cyclone by using Lagrangian method at the air mass flow rate of 2.15 g/s, which was close to the experimental condition (2.1 g/s). After loading particles for a time period of 10.5 ms, trajectories of particles for the 'I-block' inlet and 'II-block' inlet through different channels of inlet are shown in **Figure 8**. These particles were enlarged to make them visible.

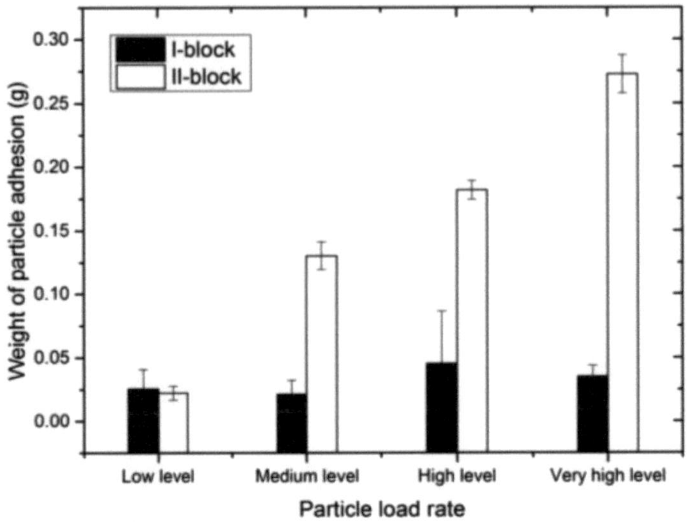

Figure 6. Weight of particle adhesion (air mass flow rate 2.1 g/s).

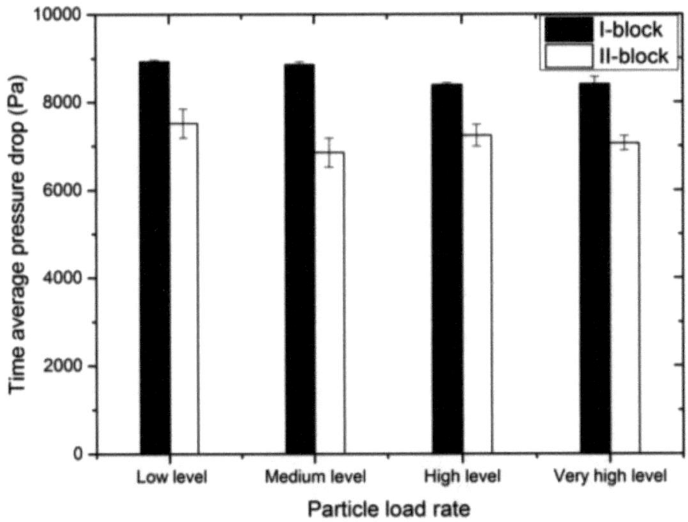

Figure 7. Time average pressure drop under different particle load rates (air mass flow rate 2.1 g/s).

It can be seen that there was a significant difference in the particle trajectories among different channels. For the 'I-block' inlet configuration, particles from the channel 'II' had quite different trajectories from that of the channel 'III' and channel 'IV'. The configuration of channel 'I', 'II', 'III' and 'IV' can be seen in **Figure 5**. There were fewer particles on the conical part of the cyclone for the channel 'II', because the majority of particles from the channel 'II' rotated under the roof that delayed their time to reach the dust collector. It is noted that the under roof region is defined as the annular volume between the roof and middle of inlet. For the 'II-block' inlet

http://dx.doi.org/10.5772/intechopen.79902

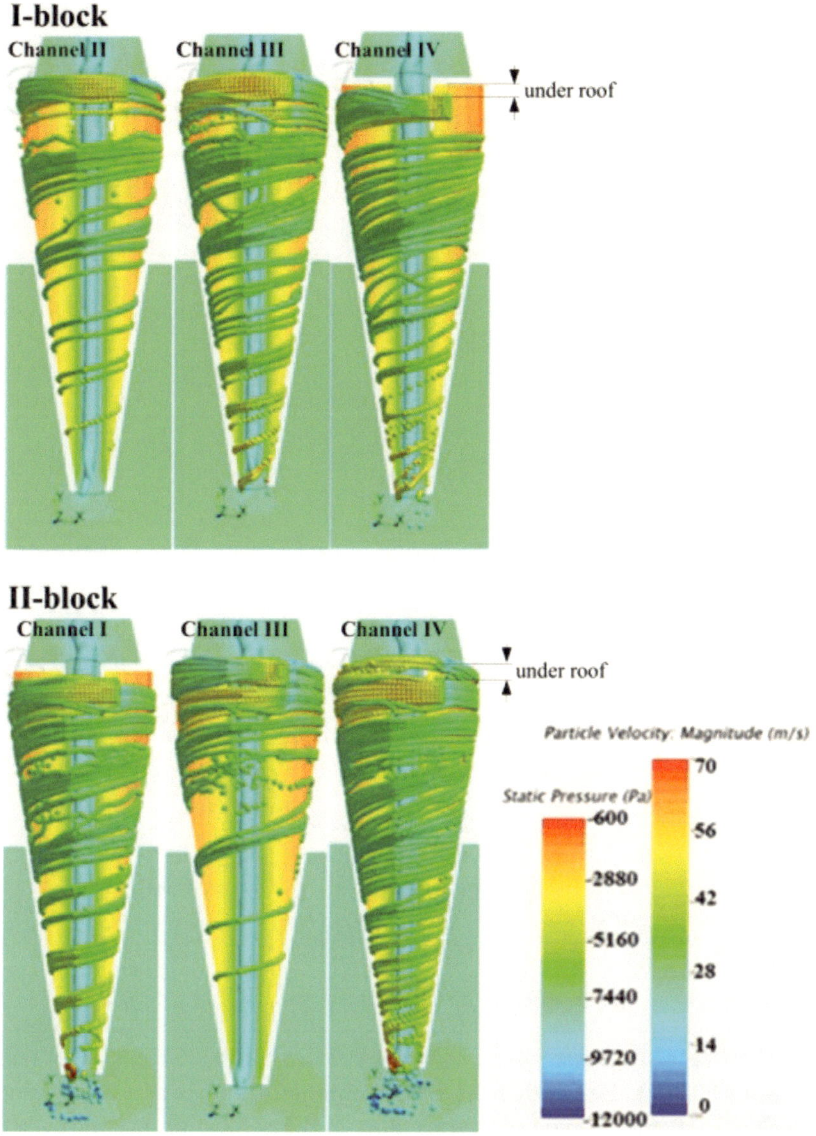

Figure 8. Trajectories of particles for the 'I-block' inlet and 'II-block' inlet (air mass flow rate 2.15 g/s).

configuration, it was because of the specific particle behaviour under the roof that the particles from the channel 'III' had quite different trajectories compared with that of the channel 'I' and channel 'IV'.

The difference between particle trajectories for different channels was believed to be determined by the secondary flow in the cyclone, as the secondary flow affected the axial transportation of particles from the inlet to the dust collector. In the central plane of the cyclone, instantaneous results of velocity vectors of the secondary flow near the inlet are illustrated in **Figure 9**.

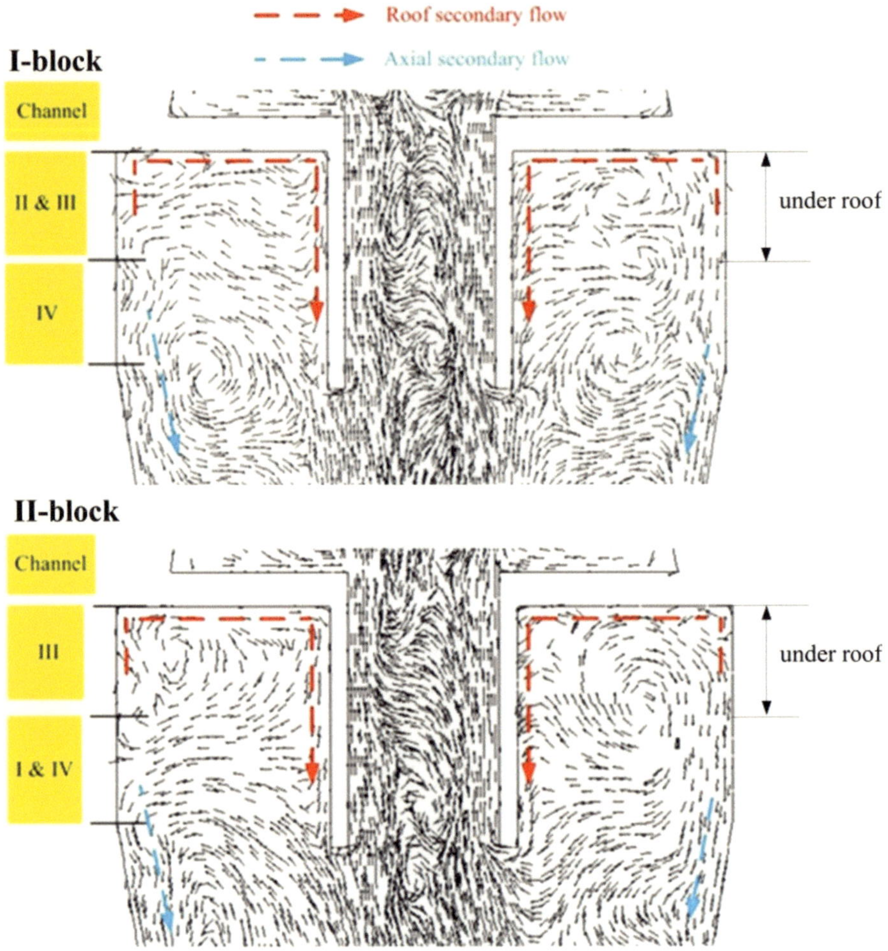

Figure 9. The secondary flow for 'I-block' inlet and 'II-block' inlet (air mass flow rate 2.15 g/s).

It can be seen that the distribution of the secondary flows for the 'I-block' inlet was similar to that of the 'II-block' inlet. For both inlet configurations, there was a roof secondary flow existing under the roof and an axial secondary flow in the conical part of the cyclone. Characteristics of the secondary flow obtained in our CFD simulation were similar to those found previously [4]. The roof secondary flow started from the upper part of inlet, rose up to the roof, moved inward along the roof, and then extended to the wall of the vortex finder, and finally entered the vortex finder via the 'lip' leakage. The axial secondary flow was driven by the pressure gradient. In the near-wall region, the axial secondary flow moved downward. In addition, it can be seen that the axial location of the roof secondary flow was the same as the axial location of the channel 'II' and channel 'III' for both inlets.

2.2.5. Discussion

The particle adhesion in the cyclone for different inlets was examined. Experimental results showed that there was a difference in the pressure drop and the particle adhesion between the two inlets. A clear correlation between the pressure drop and the weight of the particle adhesion was found. The 'I-block' inlet always had a higher pressure drop and smaller amount of particle adhesion than that of 'II-block' inlet. In the CFD simulation, it was found that the particle trajectory was clearly different between these two inlets.

Theoretically, in the cyclone, the particle adhesion was affected by various forces, including the van der Waals force, electrostatic force, aerodynamic force and the capillary force. During the experiment, as the environment condition and properties of the particle and surface were constant, the only thing that would affect the particle adhesion was the aerodynamic force. Hence, a theory was developed (also illustrated by **Figure 10**) to explain the mechanisms behind experimental results and CFD simulation results:

1. An explanation for experiment results was given. In the cyclone, firstly, the particle adhesion is negative correlated to the aerodynamic force on the particle. Secondly, since the aerodynamic force on the particle near the wall is positive correlated to the wall shear stress, the particle adhesion is negative correlated to the wall shear stress. According to the definition of wall shear stress, the wall shear stress is proportional to the free stream velocity, which is mainly affected by the swirling flow velocity on the wall of cyclone. As the swirling flow velocity in the cyclone follows the 'Rankine' vortex structure [4], the swirling flow velocity on the wall is positive correlated to the swirling flow velocity in other locations. Thus, the particle adhesion is negative correlated to the swirling flow velocity in the cyclone (more specifically, the swirling flow velocity in conical part, where particle adhesion was measured). Finally, the pressure

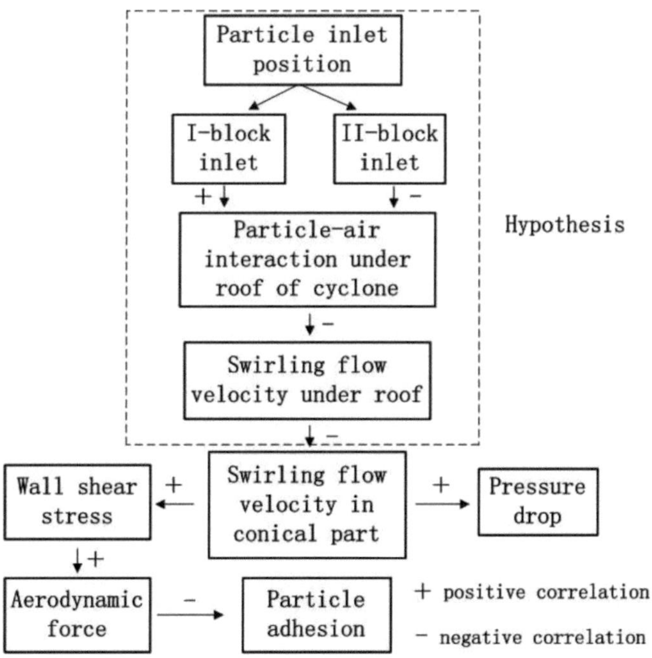

Figure 10. Theory of mechanism behind the effect of particle inlet position.

drop is positive correlated to the swirling flow velocity in the cyclone [4]. Thus, the correlation between the weight of the particle adhesion and the pressure drop was negative.

2. Based on the CFD simulation results of the velocity vectors of the secondary flow, the effect of secondary flow on particle trajectories was proposed. When the particles were trapped by the roof secondary flow, they would be lifted up by the roof secondary flow soon after they were loaded at the inlet. Then they would be carried inward in the radial direction by the roof secondary flow. However, due to the centrifugal force, the roof secondary flow was not able to carry the particles moving to the vortex finder. Therefore, the particles would stay at a radial location where the inward flow force was balanced by the centrifugal force. At the balance state, the particles would keep rotating at a radial location until the roof secondary flow was changed, resulting in fewer particles in the conical part. In contrast, when the particles were trapped by the axial secondary flow, they would have spiral trajectories and be transported to the dust collector quicker than particles trapped by the roof secondary flow. Since particles were loaded from different channels with different inlets, the interaction between the secondary flow and particle was different, resulting in different particle trajectories.

3. With the aid of the CFD simulation, a hypothesis to explain the mechanism of particle behaviour with different partly blocked inlets was proposed. As the secondary flow was not uniform around the inlet, it was believed that more particles would be affected by the roof secondary flow for the 'I-block' inlet than that of the 'II-block' inlet, because of the relative axial location between the inlet and roof secondary flow. In addition, as the particle can damp the flow, the swirling flow under the roof should be weaker for the 'I-block' inlet than that of the 'II-block' inlet. By controlling the total air mass flow rate to be constant, a weaker swirling flow under the roof would result in a stronger swirling flow in the conical part of the cyclone. As a stronger swirling flow in the conical part was linked with a higher pressure drop, the mechanism behind the effect of particle inlet position on the pressure drop was explained.

Therefore, the theory was able to explain both experimental results and CFD simulation results. However, it is noted that CFD simulation of the particle trajectory was not exactly the same as the particle trajectories in the experimental condition. In the CFD simulation, due to the high computational time, the two-way coupled particle-flow interaction and particle-particle interaction were not included. Therefore, the theory needs to be validated by tracking particle trajectory in the experiment, which is the future work.

2.3. 2D modelling of particle adhesion in cyclone

In this section, 2D modelling of particle adhesion was presented. The air flow velocity profile was generated based on our previous CFD results. The boundary layer velocity profile was linear. The transportation and adhesion of a single spherical particle from a faraway position to the wall surface in a 2D horizontal cross-section of cyclone was simulated. The influence of centrifugal force and electrostatic force on the transportation and adhesion was discussed.